Mircea A. Tamas

A VIEW OF THE CENTER

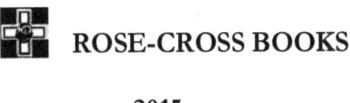 ROSE-CROSS BOOKS

2015

Copyright © 2015 by Mircea A. Tamas

www.rose-crossbooks.com

This first edition is published by **Rose-Cross Books**
TORONTO

Printed in Canada   Toronto 2015

Cover design by *Imre Szekely*

Tamas, Mircea A. (Mircea Alexandru), 1949-, author
    A view of the center : René Guénon and the traditional spirit / Mircea A. Tamas.

Includes bibliographical references.
ISBN 978-0-9865872-5-2 (paperback)

    1. Guénon, René. 2. Cosmology. 3. Symbolism. 4. Tradition (Philosophy). 5. Philosophy and religion. I. Title.

B2430.G84T356 2015   194   C2015-907077-5

TABLE OF CONTENTS

One ...... The Center

Two ...... The Center and the Spiritual Influences

Three .... The Center as a Temple

Four .... The Temple as a Center

Five .... The Vision of the Center

Six .... Solomon's Temple

Seven .... The Rock

I

THE CENTER

The Center of the World or the supreme Center of the present terrestrial humanity is called the Earthly Paradise, Agarttha, or Salem. It is the Temple of Peace (*salem*) and the House of Justice (*Beith-Din*), since the supreme Center, or any other spiritual center, which is its image, could be described as a temple (the sacerdotal aspect, corresponding to Peace) and as a palace or tribunal (the royal aspect, corresponding to Justice).[1] This Center is an image of the Heavenly Center, not as a virtual image formed by a mirror, but having full reality; also, a spiritual center is the terrestrial and visible (to human senses) image of the true Center of the World and even if the sacred orientation is done with respect to the spiritual center, it is done symbolically towards the supreme Center.

There is nothing more important, more fundamental, more essential and more plenary for a sacred doctrine than symbolizing the center, given that without the Center there is nothing.[2] As Guénon said, the Center is the origin, the source of everything, epitomized by the center of a circle, and representing the Principle's image,[3] which explains why all the

---

[1] See René Guénon, *Le Roi du Monde*, Gallimard, Paris, 1981, p. 26.
[2] A Master Mason's Lodge is "opened on the center" (Robert Macoy, *A Dictionary of Freemasonry*, Gramercy Books, 2000, p. 106).
[3] René Guénon, *Écrits pour Regnabit*, Archè, Milano, 1999, pp. 71-79. Nicholas of Cusa used the word "principle" to define the one and only God, the principle of the many gods of various people (Nicolas de Cues, *La paix de la foi*, Centre d'Études de la Renaissance, Université de Sherbrooke, 1977, p. 45).

genuine traditions and rites are so concerned with the symbolism of the center. God, through His Word, turns into the Center of the World.[1]

Meister Eckhart, in his sermon *Ez was âbent des tages*,[2] mentions the patriarch Jacob, who, "when he reached a certain place he passed the night there, since the sun had set."[3] Jacob said — Meister Eckhart stresses — "a certain place," without naming it; this place is God, the divine Being who gives life and place, being and order, to everything. In this place, the soul rests in the highest and most hidden place. This place, as Meister Eckhart suggested, is at the same time Center and God.

The Center of the World is the Principle's "location" and, in an absolute sense, there is no difference between the Center and the Principle. For example, the Lord of the World, who is, in his highest significance, a principle, identical to Manu of the Hindu tradition, personifying the cosmic Intelligence that reflects the pure spiritual Light and formulates the Law for a world or a cycle, operates and manifests his presence through the spiritual center settled in the terrestrial world.[4] The Center is *el-maqâmul-ilâhi*, the divine station where all the oppositions are solved and united, Plato's "righteous middle," the Chinese *Zhong Yong* where the Heavenly Activity operates, the "World's Navel," Aristotle's immovable mover, and, in its ultimate significance, it is indeed the very Principle, the Hindu *antaryâmi* that produces the initial impulse and then, governs and regulates the world.

The Center is the beginning, the middle and the end, where the middle is an obvious equivalent for "center" as stated in the Far-Eastern tradition (*Zhong Yong*, "the invariable middle"), but

---

[1] Guénon, *Écrits*, p. 103, René Guénon, *Symboles fondamentaux de la Science sacrée*, Gallimard, Paris, 1980, p. 110, René Guénon, *Le symbolisme de la croix*, Guy Trédaniel / Véga, Paris, 1989, p. 33.
[2] *John* 20:19.
[3] *Genesis* 28:11.
[4] Guénon, *Le Roi*, p. 13. Fabre d'Olivet already said that "Manu is the legislative Intelligence, which presides on Earth from a flood to another" (*Histoire philosophique du genre humain*, Éditions Traditionnelles, Paris, 1991, p. 238).

it also describes the point of equilibrium and harmony, of peace and immutability, obtained only as pivot of the weighing scale – the "jade balance" that represents the *Ursa Major* constellation in the Chinese tradition and is called Tulâ in Sanskrit.[1]

It is "the Pivot of the Rule" (*Dao Shu*), sheltering, like in an ark or chest, the Three Worlds (the ideogram *qu*) within the *Axis Mundi* (the ideogram *mu*).[2] As pivot, the Center is the Pole, but also the completion, the final point where all beings return; it is Alpha and Omega. Nicholas of Cusa said the same thing about God, who is identical to the Center: "God, absolute maximum, is all at once the author of all his works, the only one who knows them, and their end, since everything is in him and nothing is outside of him, who is the principle (origin), the middle and the end of everything, center and circumference of the universe"[3]; and: "the poles of the sphere coincide with the center, since the center is nothing else but the pole, that is, God."[4]

The supreme Center, like the Principle itself, has no name, no limits, escapes any definition, and it can be conceived only in a symbolic manner, as a *principial* point, without form or dimension, invisible – the image of the supreme One,[5] a point that becomes the center of a circle, or the center of a three-dimensional cross, or the center of a *swastika* – the sign of the Pole, or the hub of the wheel (the wheel of life, the wheel of law, the zodiac wheel), or the center of a flower (lotus, rose, lily, the golden sun in the center of the forget-me-not flower).

From a microcosmical perspective, the Earthly Paradise is the center of the human state, as a point generated by the intersection of the *Axis Mundi* with the domain of human possibilities, a reflected image of the universal Center. This point, identical to the Inner Palace of the Judaic Kabbalah, after

---

[1] Guénon, *Le Roi*, p. 83.
[2] Léon Wieger, *Caractères chinois*, Kuangchi Cultural Group, 2004, pp. 183, 276, 581.
[3] Nicolas de Cusa, *De la docte Ignorance*, Guy Trédaniel, Paris, 1979, p. 166.
[4] Cusa, *De la docte Ignorance*, p. 152.
[5] Guénon, *Symboles fondamentaux*, p. 84.

producing or realizing the space, makes itself the center of this space, but remains "non-located,"[1] which prompts René Guénon to refute Pascal's definition of space, "a sphere with the center everywhere and the circumference nowhere."[2] Since the central point is essentially non-located (it has no location in manifestation), the *principial* center is, in fact, nowhere, and the circumference, or all manifested and produced beings,[3] is everywhere.[4]

Guénon refers to the absolute Center and not to its substitutes; with respect to the manifestation, Nicholas of Cusa stated: "the world machine will have, one might say, its center everywhere and its circumference nowhere."[5] Applying the metaphysical principles to our world, Cusanus highlighted the error of geocentrism and explained that Earth is not the center of the universe, that any celestial body is, apparently, the center of the world for its inhabitants, and so, the manifested world has the center everywhere, since a universe that changes continuously could not have a fixed and immobile center.[6] In the same manner, any secondary genuine and orthodox tradition, regularly[7] derived from the Primordial Tradition, considers its spiritual center as "the center."

"The hidden point" – the immutable Center, the immovable mover (*to prôton kinoun akinêton on*), the pivot – does not belong to the universal manifestation, as it is the principle of the

---

[1] Guénon, *Écrits*, pp. 173-9.
[2] René Guénon, *Le symbolisme de la croix*, p. 148. This definition is used by the ritual of *The Holy Royal Arch of Jerusalem* (See W. L. Wilmshurst, *The Meaning of Masonry*, Gramercy Books, New York, 1980, p. 74).
[3] We relate "being" with "to be."
[4] Guénon, *Le symbolisme de la croix*, p. 151.
[5] Cusa, *De la docte Ignorance*, p. 155.
[6] "It is impossible for the world machine to have this sensible earth, air, fire, or anything else for a fixed and immovable center. For in motion there is simply no minimum, such as a fixed center. (...) And although the world is not Infinite, it cannot be conceived of as finite, since it lacks boundaries within which it is enclosed. (...) Therefore, just as the earth is not the center of the world, so the sphere of fixed stars is not its circumference" (Cusa, *De la docte Ignorance*, pp. 150-151).
[7] That is, following the sacred rules or regulations.

universe "beyond" the universe, and therefore non-manifested. The *principial* point's "modes" are the ones located in manifestation; they are the worldly knots, "the vital centers" vibrating by reflection the primeval vibration.

On the other hand, the circumference, as the World's limit, is nowhere, considering how the universal manifestation, for our human mind, extends indefinitely. Such an interpretation fails to see beyond the *prakritian* perspective,[1] but Nicholas of Cusa, fortunately, does not stop here. Cusanus, in fact, has revitalized a twelfth century hermetic adage, which stated that God is a circle with the center everywhere and the circumference nowhere, and Pascal, resuming Cusanus' formula, two centuries later, condemned it to the level of the world, which, obviously, pleased modern mentality. From a metaphysical point of view, as we have already mentioned, Guénon explained how the formula should be rewritten according to inverse analogy and based on a Daoist text, transmitted by Zhuang Zi: "The point, which is the Pivot of the Rule (*Dao Shu*), is the immovable center of a circumference on which all the contingencies, distinctions and individualities rotate."[2]

Similarly, a circle could symbolize the Biblical syntagm "The Being is the Being" (*Eheieh asher Eheieh*),[3] in which case the central point is the Logos and the circumference is the manifested world and the geometrical locus of the *principial* point's "modes" – worldly knots. In such a representation, the circumference's ubiquity obviously designates the universal Existence (the circumference is everywhere), which appears as a worldly reflection of the *principial* omnipresence of the hidden point, of the Center that does not belong to the universal manifestation (the center is nowhere); or, in Nicholas of Cusa's

---

[1] See our work, *About the Yi Jing*, Rose-Cross Books, Toronto, 2006, p. 96.
[2] Léon Wieger, *Les pères du système taoïste*, Les Belles Lettres, Paris, 1950, p. 219 (*Tchoang-tzeu*, chap. 2, C).
[3] Guénon, *Le symbolisme de la croix*, p. 102, and related to Cusa's statement, "God is" (*De la docte Ignorance*, p. 148).

language, we could say that God is everywhere as explication and nowhere as complication.

Anticipating the equivocal effect, Cusanus added a remarkable clarification to the formula: "Therefore, the world machine will have, one might say, its center everywhere and its circumference nowhere, for its circumference and center is God, who is everywhere and nowhere."[1] To illustrate this fundamental statement, Nicholas of Cusa uses the mathematical theory of limits, introducing the two "infinites" (which Pascal hijacked and distorted later), and applies the concept of movement – a basic characteristic of the universe, since only the Principle is the immovable mover – to the geometrical diagram mentioned above, marking the limits or the extremes: the movement's minimum in the immovable center ("the infinitely small") and the maximum on the circumference ("the infinitely large"), extremes that coincide "at infinity," namely "beyond" the universal manifestation, which reveals God – the One and only "located" at infinity – as maximum and minimum, center and circumference, since only in God do the two infinites coincide.

All the traditional doctrines agree on this. In the Islamic tradition, the Throne (*El-Arsh*) is all together center and circumference; and Angelus Silesius stated: "God is my circle and my center-point." In the Hindu tradition, "Âtmâ is smaller than a grain of paddy, than a barley corn, than a mustard seed, than a grain of millet or than the kernel of a grain of millet. Âtmâ is greater than Earth, greater than Atmosphere, greater than Heaven, greater than all Three Worlds"[2]; "Those who know beyond this the supreme Brahma, the vast, hidden in the bodies of all creatures [the center], and alone enveloping everything [the circumference], as the Lord, they become immortal."[3] In the Christian tradition, "the Kingdom of Heaven is like to a grain of mustard seed, which a man took, and sowed in his field; which indeed is the least of all seeds: but when it is

---

[1] Cusa, *De la docte Ignorance*, p. 155.
[2] *Chândogya Upanishad*, 3.14.3.
[3] *Svetâshvatara Up.*, 3.7.

grown, it is the greatest among herbs, and becometh a tree, so that the birds of the air come and lodge in the branches thereof."¹ Thus, God is the "minimum" – the center, the *principial* point that does not take any space, since it is the space's principle, the hidden point that is nowhere, but which is everywhere by means of the "knots" it illuminates with its primordial vibration, and the "maximum" – the circumference that encompasses everything and is reflected everywhere as manifestation and multiplicity, but since it is beyond our touch, it is nowhere. Zhuang Zi said: "The Principle is in all the beings [as "minimum"]. Therefore it is called great, supreme, universal, complete, all [as "maximum"]."²

It is most interesting to see the great spiritual master Ibn 'Arabî applying this formula to Abraham, when he says "the bosom friend (*al khalîl*) was designated by this name because he penetrates³ and encompasses everything that qualifies the divine Essence,"⁴ which indicates the special function of Abraham, not only for the three religions but also from an esoteric point of view (René Guénon said: "For those who know the Islamic tradition, it is easy to understand the relation between Abraham and Masonry"⁵).

Nicholas of Cusa's formula tries to surpass the "machine" as world, rising to the *principial* level, as another text confirms: "God is within everything, because He is an infinite center; without everything, because He is an infinite circle; penetrating everything, because He is of an infinite diameter, the principle of everything being the center, the end of everything being the circumference, the middle of everything being the diameter, efficient cause being the center, formal cause being the diameter, the terminal cause being the circle, the center

---

¹ *Matthew* 13:31-32.
² *Tchoang-tzeu*, chap. 13, G, Wieger, *Les pères*, p. 315.
³ *Takhallala*, "to penetrate," derives from the same verbal root as *khalîl*.
⁴ Ibn 'Arabî, *Le livre des chatons des sagesses*, Les Éditions Al-Bouraq, Beirut, 1997, p. 167, tr. Charles-André Gilis.
⁵ René Guénon, *Études sur la Franc-Maçonnerie et le Compagnonnage*, Éditions Traditionnelles, Paris, 1980, II, p. 165.

producing the being, the diameter ruling, and the circle keeping."[1] But Cusanus does not forget the symbolism of the center: "the center of the maximum sphere is equal to the diameter and to the circumference. The infinite sphere[2] is equal to its center, but even more, the center is all this: length, width, depth[3]; it is the maximum, simple and infinite. The center precedes all length, width and depth, and is also the end and the middle of all this."[4] "God is the Center of the World, and He is the center of the Earth and of all the other spheres, and of everything in this world, He who is at the same time infinite circumference of everything."[5]

Ezekiel, whose vision of the Temple makes him a prophet of the center, had also a vision of God as universal vortex, a "world machine" of fire and lightning: "And I looked, and, behold, a whirlwind came out of the north, a great cloud, and a fire infolding itself, and a brightness was about it (…) Also out of the midst thereof came the likeness of four living creatures. (…) And every one had four faces, and every one had four wings. (…) Their wings were joined one to another; they turned not when they went; they went every one straight forward. As for the likeness of their faces, they four had the face of a man, and the face of a lion, on the right side: and they four had the face of an ox on the left side; they four also had the face of an eagle. (…) And the living creatures ran and returned as the appearance of a flash of lightning. Now as I beheld the living creatures, behold one wheel upon the earth by the living creatures, with his four faces. (…) their appearance and their work was as it were a wheel in the middle of a wheel. (…) As for their rings, they were so high that they were dreadful; and their rings were full of eyes round about them four. And when the living creatures went, the wheels went by them: and when

---

[1] Cusa, *De la docte Ignorance*, p. 84. For Cusa, Jesus Christ, "the maximum man," is the intellectual nature's center and circumference (pp. 184, 202).
[2] The infinite sphere should be compared to Guénon's "universal vortex."
[3] The three-dimensional cross.
[4] Cusa, *De la docte Ignorance*, p. 88.
[5] Cusa, *De la docte Ignorance*, p. 152.

the living creatures were lifted up from the earth, the wheels were lifted up."[1]

We must admit that Ezekiel's description is impressive, this "cosmic chariot" stimulating modern minds into suggesting all kinds of fantastic hypotheses. We notice the vehicle's harmony, its perfect isotropy due to the spherical wheels, while the four creatures with their four identical faces appear united in a complete wholeness, enveloped by fire and light. Ezekiel's vision represents, we may say, the primordial luminous sphere with its length, width, and depth delineating the three-dimensional cross, yet it also suggests a spatial *swastika* – symbol of the Pole, which has on a horizontal plane (as it is usually drawn) four branches associated with the four creatures. In the Christian tradition, a similar figure is the *gammadia* or "the cross voided throughout," composed of four Greek letters *gamma*, where the inner void in the shape of a cross symbolizes Jesus Christ and the four *gammas* designate the four Evangelists.[2] Christ and the four creatures or the four Evangelists, the Prophet Muhammad and the four Kolaphâ, Horus and his four sons correspond either to the center and the four corners or to the center and the four cardinal points, in the latter case, like in the case of the *swastika*, the circumference being replaced by the two perpendicular diameters that underline "the orientation" and define the whole manifestation; the center and the four corners correspond to the "five points" that traditionally determine the location of a temple.[3] These "five points" are not only the essential summary of the circle (center and circumference), but they define an intersection of the universal vortex with a plane as a degree of the universal Existence, with the center as an image of the supreme and absolute Center.

The Center sends its reflections into the various worlds and into the various cosmic cycles, including our world, each reflection or image, for a cycle (major or minor) or a world, standing for the essential kernel (*luz*), the spiritual marrow

---

[1] *Ezekiel* 1:4-19.
[2] Guénon, *Le symbolisme de la croix*, p. 71.
[3] Guénon, *Symboles fondamentaux*, pp. 299-300.

where the divine presence finds a dwelling, where the Heavenly Activity operates, producing and supporting that particular state of existence, world or cycle, "here" and "now." This spiritual center is the keeper of the sum of all sacred knowledge needed to govern that world; it is the residence of the Primordial Tradition revealed through sound and light to the Lord of the World, a revelation understood as an active participation of the Lord of the World to the Heavenly Mysteries. From this primordial and supreme center of a world, the Tradition overflows, like the paradisiacal rivers, irrigating the whole world, generating secondary centers and traditions, which can exist simultaneously or in a sequence.

The Center of the World – we have to understand that – is invisible, inaccessible, untouchable, belonging to the sacred geography, without a specific "localization" (as the temple of the Rose-Cross is) and only by particularization to a world or a specific civilization becomes "materialized"; but, since the whole nature is a symbol of the supernatural,[1] this "materialization" encompasses an extremely rich symbolism, which means that any spiritual center could be a temple, a palace, a monastery, a garden, a forest, a city, a mountain, an island, a cavern, a fountain, a tree, or a hearth.[2] For terrestrial humanity, the spiritual center or the world's heart was at the beginning of the Golden Age synonymous with the Earthly Paradise, and at the end of the Iron Age, the Heavenly Jerusalem will descend and mark it again.

However, this rich symbolism does not imply that a temple, a sacred stone or any other "central" mark is just symbol with merely virtual or theoretical authority, which would infer that our world could exist by itself – a pure impossibility; but, they are, each of them, the "house of God," the well-lit and well-regulated location[3] where the divine presence operates, a location that has to have special attributes or, in a deeper sense, no attributes at all, a location that, consequently, has to be

---

[1] Guénon, *Écrits*, p. 57.
[2] See Guénon, *Symboles fondamentaux*, p. 109.
[3] Guénon, *Le Roi*, p. 23, *La Grande Triade*, Gallimard, 1980, p. 139.

ritually prepared to give God a home, and so, we are not talking here about any pantheism, this absurd modern invention.

Traditional symbolism compares the Center with the human heart – the center of the being and "divine residence" (the Hindu *Brahma-pura*).[1] In the Hindu tradition, *Âtmâ* dwells in the heart; in the Islamic tradition, the heart is Allâh's throne; in fact, all the genuine traditions consider the heart as a "house of God." Yet only virtually is each heart a divine residence, since the soul has to be prepared to receive God's presence, and for Meister Eckhart that is the meaning of the merchants' expulsion from the Temple. In his sermon, *Intravit Iesus in templum et coepit eicere vendentes et ementes*,[2] Meister Eckhart regards the Temple as the human soul, and God, Who created and formed the soul after His resemblance, wants this temple to be empty, that is, only when this temple is liberated of all obstacles, will God come and reside in it. Al-Hallâj stated the same thing: "When Allâh chooses a heart, He empties it of all that is not Him."

In a similar way, a temple, a tree, or a stone must be prepared and emptied of all the obstacles and "bad" influences and must be purified in an effective manner to become the house of God. In addition, since the Center is "non-located," the location[3] for the spiritual centers established in the world is an imperative element connected to sacred geography and the science of orientation, all these centers, which are images of the Center, showing similar topographic characteristics and obeying the laws that govern the activity of the spiritual influences.[4] Directly or indirectly, God indicates where the location should be, as God indicates the architectural plans for the center, given that a spiritual center has nothing to do with the individual order and everything with the divine order, because the house of God must be sacred within and without. Therefore, in various traditions, a divine animal guides the founder of a center, like in the case of Thebes, when Cadmus (that is, "the

---

[1] Guénon, *Le Roi*, p. 25.
[2] *Matthew* 21:12.
[3] We prefer the word "location," bearing in mind its relation to Sanskrit *loka*.
[4] Guénon, *Le Roi*, p. 37.

primordial," an image of Adam Kadmon) listened to the Delphic oracle and followed a sacred cow (with a half moon on her flank). Thebes – Guénon stressed – is a name designating the spiritual centers, in view of the fact that *Thebah* is the Hebrew name of Noah's Ark, which is as well a representation of the supreme center[1]; also, Guénon added, the foundation of a city could symbolize the institution of a doctrine or of a traditional form, and Thebes is a good example, since Amphion built it using the music of his lyre, an important instrument in Orphism and Pythagorism (related to the science of rhythm).[2]

"And he dreamed, and behold a ladder set up on the earth, and the top of it reached to Heaven: and behold the angels of God ascending and descending on it. And, behold, the Lord stood above it, and said, I am the Lord, God of Abraham thy father (…) I am with thee, and will keep thee in all places whither thou goest, and will bring thee again into this land; for I will not leave thee, until I have done that which I have spoken to thee of. And Jacob awaked out of his sleep, and he said, Surely the Lord is in this place; and I knew it not. And he was afraid, and said, How dreadful is this place! this is none other but the house of God, and this is the gate of Heaven. And Jacob rose up early in the morning, and took the stone that he had put for his pillows, and set it up for a pillar, and poured oil upon the top of it. And he called the name of that place Bethel: but the name of that city was called Luz at the first."[3]

In the case of Jacob, God indicates, in a dream, the location of the spiritual center, marked by a stone that is the house of God and, very importantly, the gate of Heaven (the center is the contact point between Heaven and Earth[4]); moreover, Jacob

---

[1] Guénon, *Le Roi*, p. 91.
[2] Guénon, *Le Roi*, pp. 89-90.
[3] *Genesis* 28:12-19.
[4] Guénon, *Écrits*, p. 112. Meister Eckhart, in his sermon *Jêsus hiez sîne jüngern ûfgân in ein schiffelîn und hiez sie varn über die wuot* (*Matthew* 14:22), said, quoting St. Augustine, that the soul is a point between time and eternity. In the Hindu tradition, a *tirtha* ("holy place") is considered the isthmus or the ford between this world and the divine one (identical to the Islamic *barzakh*); through the

accomplishes the rite of unction, purifying and preparing the stone to be the support for the spiritual influences. Before Jacob, Abraham established, guided by God, spiritual centers: "And the Lord appeared unto Abram, and said, Unto thy seed will I give this land: and there builded he an altar unto the Lord, who appeared unto him. And he removed from thence unto a mountain on the east of Bethel, and pitched his tent, having Bethel on the west and Hai on the east: and there he built an altar unto the Lord, and called upon the name of the Lord."[1] "Then Abram removed his tent, and came and dwelt in the plain of Mamre, which is in Hebron, and built there an altar unto the Lord."[2] "And they came to the place which God had told him of; and Abraham built an altar there, and laid the wood in order, and bound Isaac his son, and laid him on the altar upon the wood."[3]

In the Islamic tradition, Abraham is not only the founder of a spiritual center, but also of a holy land. As previously stated, the central point produces (or realizes) the space, and makes itself the center of this space, but without "location"; yet, for our human mind space is an indispensable condition of existence, and so is the time, the Judaic Kabbalah describing therefore the center of the world as the center of space and time.[4] The Earthly Paradise is portrayed not as a central point without dimension, but as a garden, that is, as holy land or holy space, even though the Tree of Life marks the center. In the Hindu tradition, the *ashwamedha* ("the horse sacrifice"), an essential Vedic ritual, embodies the production of the holy space, which, at the highest level, symbolizes the sacralization of the land as the spiritual influences of the center (the sacred horse, similar to Cadmus' cow) propagate in the form of three-dimensional waves.

---

tirtha the being ascends toward Liberation and the spiritual influences come into the world.
[1] *Genesis* 12:7-8.
[2] *Genesis* 13:18.
[3] *Genesis* 22:9.
[4] Guénon, *Écrits*, p. 100.

A center is an emanation or a reflection of the supreme spiritual center (the center of the Primordial Tradition), and hence, it is an image of this one and virtually identical to it,[1] which explains why the secondary centers are described in a similar manner[2]; any region that surrounds such a spiritual center is, because of that, a "holy land." All traditions affirm – René Guénon concluded at the end of his *Le Roi du Monde* – the existence of an archetypal Holy Land, the prototype for all other "holy lands," the spiritual center to which all others are subordinate. The Holy Land is "non-located" and different traditions situate it in the "invisible world," but there are allusions to some specific locations in some specific regions and Guénon asked rhetorically if we should take it literally or symbolically, or whether it is both at the same time. The simple answer is (Guénon explained) that both geographical and historical facts possess a symbolic validity that in no way detracts from their being facts, but that actually, beyond this obvious reality, gives them a higher significance.[3] This archetypal Holy Land is "the supreme region" or "the supreme space," the meaning of the Sanskrit term *Paradêsha*, which the Westerners adopted as "Paradise," and, indeed, Earthly Paradise is the supreme Center of the World, Holy Land and Heart of the World, even though other traditions called it Tula, Luz, Salem, or Agarttha.[4]

For the Judaic and Christian traditions, the holy land is the one promised to Abraham and located where Jerusalem is; however, since all the secondary centers, founded in order to adapt the Primordial Tradition to different spatial and temporal conditions, are images of the supreme center, Jerusalem could be an image of Salem, and virtually identical to it, as Sion, even

---

[1] Guénon, *La Grande Triade*, p. 138.
[2] Guénon, *Le Roi*, p. 39.
[3] Guénon, *Le Roi*, pp. 95-96.
[4] Guénon, *Symboles fondamentaux*, pp. 108-109. "The word Salem has never described a town but, taken as the symbolic name of Melchizedek's residence, can be considered the equivalent of the term Agarttha" (Guénon, *Le Roi*, p. 49).

though a secondary center, is symbolically equal to the supreme center, and the holy land is not only the land of Israel, but a substitute for the Earthly Paradise.[1] As Vulliaud noted, "the Tabernacle of the Holiness of Jehovah, the residence of Shekinah, is the Saint of Saints, which is the heart of the Temple, which is itself the center of Sion (Jerusalem), as the Holy Sion is the center of the Land of Israel, and as the Land of Israel is the center of the world."[2] There is nothing unusual here, since if we refer to the *ashwamedha* ritual we understand that in a normal society the whole world should be a holy land emanating from the center, with the specification that a symbolic hierarchy of the spiritual Pole's successive approximations should be observed, like in the case of the Judaic Kabbalah or of some fairy tales, where the sacred "powers" are hidden, for example, inside two flies, hiding in a duck, hiding in a rabbit, hiding in a bear, hiding in a cave. Guénon went on further to mark the center even more rigorous, by considering the Ark of the Covenant in the Tabernacle and on the Ark itself the point between the two Cherubim where the *Shekinah* manifests, marking the "spiritual Pole."[3] Starting from this Pole, we notice a series of extensions gradually assigned to the notion of center, the appellation "Center of the World" or "Heart of the World" being legitimately applied to each of these extensions.[4]

The *Tanakh* refers to "God who dwells between the Cherubim," which is traditionally interpreted as the seat of the *Shekinah* (it is the Mercy seat of the *Bible*: "the Cherubim shall stretch forth their wings on high, covering the Mercy seat with their wings"[5]); in the *Book of Genesis*, the Cherubim are described guarding the way to the Tree of Life, that is, to the Center. The Cherub should be related to the Babylonian Karibu, placed at entrances to palaces or temples as guardian, that is, at the

---

[1] Guénon, *Le Roi*, p. 57.
[2] Paul Vulliaud, *La Kabbale Juive*, Émile Nourry, Paris, 1923, I, p. 509.
[3] Guénon, *Le Roi*, p. 56, *Écrits*, p. 112.
[4] Guénon, *Symboles fondamentaux*, p. 106.
[5] *Exodus* 25:20.

entrance to the center, and we may compare the two Cherubim to the pillars of Solomon's Temple, even though they have, besides the guardian function, an intermediary role, like *Shekinah* herself, related to the spiritual influences.

As the Cherubim guard the Earthly Paradise (that is, the supreme Holy Land),[1] so the pillars guarded Solomon's Temple and, inside it, in the Holy of Holies, the golden cherubim made of olive wood guarded the Ark of the Covenant, in the same way the Templars guarded the holy land where Jerusalem was situated. The particular "holy lands" are as many as the particular traditional forms,[2] and so are the guardians, therefore the Druzes and the Assassins (or Ismailians) – Guénon specified – were called also "the guardians of the holy land," a different holy land than Palestine.[3] Each holy land is the spiritual center of an orthodox traditional form, and other people, not only the Israelites, have been in possession of a sacred space with a spiritual center, which had for them a role comparable to the one the Temple of Jerusalem had for the Jews.[4]

The series of extensions gradually assigned to the notion of center, as Vulliaud and René Guénon pointed them out for the Judaic tradition, are present in the Islamic tradition too, where, it is said, the holy location of Kaaba was established before any other part of the earth; Allâh simply extended the earth around Kaaba in a series of concentric circles, Becca, Mecca, the *haram*, and, eventually, the rest of the earth, and so we notice the radiation of the spiritual influences from the center to a series of circumferences, a center that is "the mother of all cities" (*umm el qurâ*), "the center of the world" (*wast ed dunya*) and "the navel of the earth" (*surrat el ardh*).[5]

---

[1] *Genesis* 3:24.
[2] Guénon, *Symboles fondamentaux*, p. 108.
[3] Guénon, *Symboles fondamentaux*, p. 105.
[4] Guénon, *Symboles fondamentaux*, p. 107.
[5] Gaudefroy-Demombynes, *Le pèlerinage à la Mekke*, Paul Geuthner, Paris, 1923, p. 30.

Regarding the last expression, we note that René Guénon, in *Le Roi du Monde*, dedicated a whole chapter to this symbol of the center. He said that *omphalos* (in Greek "navel") was one of the most remarkable and widely spread symbols[1]; the best-known *omphalos* was the one in the temple of Delphi, symbolizing the spiritual center for all ancient Greece.[2] The material representation of *omphalos* usually consisted of a sacred stone, a *betyl*, that is, a "house of God" (Hebrew *Beith-El*), and this stone was the true "divine habitation" (*mishkan*), the seat of the *Shekinah*, a designation that was later used to name the Tabernacle.[3]

In the Islamic tradition, the sacred stone also had an essential role. When Abraham placed the Black Stone (*Al-hajar Al-aswad*) into the corner of Kaaba, a bright flame sprang up and spread, reaching the limits of the *haram*. The demons that kept Abraham under surveillance were struck by the light and were stopped precisely at the landmarks Abraham raised to guard the sanctuary. As the Prophet Muhammad said, the *haram* or the holy land was established at the same time as Heaven and Earth, and the angels sent by Allâh to protect Adam against Satan took their positions precisely by the stones erected as landmarks, which became the limits of the *haram*; the angel Gabriel (Jibrail) taught Abraham the exact location of these landmarks.[4]

From the spiritual center (Kaaba, the Black Stone) the divine presence spread as tongues of fire (similar to the Holy Spirit of the Christian tradition) generating the holy space, the *haram*, and the stones marking the limits of the holy land are projections of the Black Stone in much the same way the points of a circumference are projections of the center.[5]

---

[1] *Le Roi*, p. 74.
[2] *Le Roi*, p. 76, *Écrits*, pp. 90-1.
[3] *Le Roi*, p. 77. It is very important to note what Guénon added here, in an elusive manner, that "all this relates naturally to the subject of 'spiritual influences' (*berakoth*)."
[4] Gaudefroy-Demombynes 24.
[5] Philostratus wrote: "Thirty stades from this river, they met altars with these inscriptions: To my father Ammon. To my brother Hercules. To Minerva Providence. To Jupiter the Olympian. To Cabires of Samothrace. To Indian

One of the most famous symbolic circumferences is the Zodiac. Each of the concentric circumferences that radiates from the central point has a zodiacal function and could be a zodiacal image, but the Zodiac corresponds in the first place to the boundaries of the holy land as "the frame of the Cosmos," which is its most customary meaning. However, as previously explained with respect to the symbolic hierarchy of the spiritual Pole's successive approximations, the same symbol used to describe the Center could be extended to the Temple, to the Holy City, to the Holy Land, and, at the limit, to the Holy World. René Guénon said: "the zodiacal constitution is to be found as a general representation of the spiritual centers belonging to different traditional forms."[1]

The word "temple" comes from Greek *temno*, "to cut"; hence, Greek *temenos* "a piece of land cut off" and "a piece of land sacred to a god"; hence, Latin *templum*, "a sacred place." Cutting off a piece of land means actually to separate somehow this portion from the outside land. The outside land becomes in this case the "outside darkness," the wild woods, a profane and tenebrous place. The land inside the cut becomes the sacred kernel, the self-illuminated place, the "temple," the holy land. In Latin, we find an Indo-European word *seco*, "to cut"; its root provided the words *sacer*, "saint, sacred," *sacrificium*, "sacrifice, immolation."[2] A "cutting" is the circular furrow dug by Romulus at the foundation of Rome; a "cutting" is also the wall, its primitive function being to protect and defend, not so much a physical protection, but a spiritual one, similar to the Cherubim's role of protecting the Garden of Eden.

---

sun. To Apollo of Delphi. They also saw a bronze stela with the inscription: Here Alexander stopped. It is believed that the altars were built by Alexander, jealous to mark in this glorious way the limits of his empire" (Philostratus, *Apollonius of Tyana*, II, 43).

[1] Guénon, *Symboles fondamentaux*, p. 115. After d'Alveydre, the highest circle and closest to the mysterious center Agarttha is composed of twelve members and corresponds to the Zodiacal zone. "Celebrating the magical mysteries, they carry the hieroglyphs of the Zodiacal signs" (Saint-Yves d'Alveydre, *Mission de l'Inde en Europe*, Dorbon, 1949, p. 34).

[2] *The Everlasting Sacred Kernel*, p. 61.

The most enigmatic and famous example of a protected city is Atlantis. Beyond the historical debate about the existence and location of Atlantis, Plato's description perfectly matches the symbolical picture of the Center of the World. The mountain and the island are the most common symbols for a spiritual center. Atlantis is built on a mountain located in the center of an island. The mountain is the *Axis Mundi*, the Hindu *Mêru*. Plato tells us that Poseidon, god of the Sea, to protect Atlantis, enclosed the city with two circular ramparts and three circular fosses, alternating land and water.[1] Atlantis became inaccessible; no man could get to the spiritual center.[2] Poseidon built the enclosures, playing the role of the Great Architect of the Universe: the sacred "cutting" is so important that the gods themselves are involved as Masons, and build the divine enclosure.

Therefore the Zodiac is the celestial archetype of the terrestrial town,[3] and it encloses the Universe as the "cutting" encloses the "sacred place." In Masonry, the "chain of union," which surrounds the Masonic Lodge, usually has twelve knots, being a reflection of the Zodiac[4]; the "chain of union" encloses and keeps together the elements of the Lodge, symbolizing the Zodiac framing the Cosmos.

In the Christian tradition, *The Book of Revelation* presents the end of the world and the birth of a new one in the following manner: "I saw a new heaven and a new earth; the first heaven and the first earth had disappeared now, and there was no longer any sea. I saw the holy city, and the New Jerusalem,

---

[1] The Tibetan tradition describes the Center as follows: "In the center was a great mountain, Rirab Lhunpo, a four-sided column of precious stones, the abode of gods. Around it lay a lake, and around the lake a circle of golden mountains. Beyond the golden mountains was another lake, encircled again in turn. In all there were seven lakes and seven rings of golden mountains, the innermost being the mightiest. Seven times earth, seven times water" (Thubten Jigme Norbu, *Tibet*, Simon and Schuster, New York, 1970, pp. 19-20).

[2] Plato, *Critias* 113.

[3] Guénon, *Symboles fondamentaux*, p. 121.

[4] Guénon, *Symboles fondamentaux*, p. 388.

coming down from God out of heaven, as beautiful as a bride."[1] The Heavenly Jerusalem is described as having twelve gates, obviously a symbol of the Zodiac. The production of a new Cosmos is symbolized by the foundation of a "sacred place," of a spiritual center, of a holy city with a zodiacal frame. The Zodiac keeps together and orders all the elements of the Cosmos, from the most luminous to the most tenebrous. The countless elements or knots are reflections of the Center of the World, of the supreme City, *Brahmapura*, and so they can be represented in their turn as enclosures or cities.

As we see, the "cutting" represents not only a protection against the evil forces or an obstacle against ignorance, but also a frame to sustain and order the elements of the enclosure. Any alteration of the sacred frame would shatter the harmony, the balance, and the peace, leading to chaos and the devils' invasion. When Phaeton, Helios's son, drove his father's chariot, the solar vehicle, he could not follow the zodiacal path, and his "trespassing" destroyed the order, burning the world. Remus, Romulus' twin brother, was punished and killed because he jumped over the sacred "cutting" of Rome.[2] After that, a Roman rule stipulated the death penalty for the soldier who crossed the wall of the camp instead of using the gate.[3]

The safe place for crossing is the gate, but the gate is also very well guarded: cherubs, lions, dragons, monsters or sphinxes are the *dvârapâlas*, "the guardians of the door," punishing the intruders and testing the "elected" ones. The gate is itself an image of the center (Jacob's "gate of Heaven"), as explicitly illustrated by the Medieval cathedrals where the Zodiac envelops the gate,[4] materializing the spiritual symbolism of Jesus, the Gate and the Center, surrounded by the twelve

---

[1] *Revelation* 21:1-3.
[2] A sacrifice always accompanies the foundation of a center. Trophonius, the legendary architect who built Apollo's temple at Delphi, beheaded his brother Agamedes before establishing the oracle in his famous cave at Lebadeia.
[3] Jackson Knight, *Vergil, Epic and Anthropology*, Barnes & Noble Inc., 1967, p. 219.
[4] Jean Hani, *Le Symbolisme du Temple Chrétien*, Guy Trédaniel, 1978, p. 94.

Apostles. Here, the twelve disciples do not represent the Zodiac as the frame of the Cosmos, not even as the frame of the holy land; they symbolize "the inner circle," the spiritual influences sent from the center into the world to make it holy. They are the twelve windows of the *Ming-Tang*,[1] the twelve gates of Heavenly Jerusalem, which are the "eyes" of the Center that measure with their vision the whole existence. In a similar manner, the return to the Center means the return of the twelve rays to the central point and there is an allusion to this symbolism in the Judaic tradition, when Moses sent, obeying God's order, twelve men to spy out the land of Canaan.[2] We should mention, in relation to the Zodiac, that the twelve spies represented the twelve tribes of Israel, and here we find again the same symbolism exposed when the children of Israel's encampment is described: the Tabernacle in the center, "the Levites shall pitch round about the Tabernacle," guarding it (the inner circle), and the twelve tribes shall pitch parted into four, three tribes for each cardinal point (the outer zodiacal circle).[3]

In the case of a holy land, the difference between the center and the zodiacal points of the circumference, or between Kaaba and the *haram's* border stones, is the same as the difference between *Shekinah* and the Cherubim, or between spiritual authority and temporal power. As Guénon said, the keepers of the Tradition and the distributors of the doctrine are situated in the center, while the guardians, the defenders, and "the external cover" of the center are situated at the limits of the holy land, this role of guardian and defender being the natural function of the *Kshatriyas*; therefore, in the Christian tradition, it was normal to have the Orders of Chivalry as guardians of the holy land, and it was also normal to have the Temple of Solomon as symbol, although for the Templars, "guardians of the holy

---

[1] Guénon, *Triade*, p. 141, Marcel Granet, *La pensée chinoise*, Albin Michel, Paris, 1980, pp. 150-151, Luc Benoist, *Art du Monde*, Gallimard, Paris, 1941, p. 90.
[2] *Numbers* 13:17.
[3] The twelve tribes and Ezekiel's four animals (in relation to the four cardinal points) are included in the ritual of *The Holy Royal Arch* (Wilmshurst 155-6).

land," Solomon's Temple was not the material edifice, but the ideal one, as symbol of the supreme Center.[1]

Undoubtedly, during the present *Manvantara*, there were various holy lands situated in different conditions of space and time. Yet, in each case, for a holy land to be holy it was mandatory to shelter the divine presence; in the Judaic tradition, the divine presence is called *Shekinah* and the Tabernacle is considered the "habitacle of God" or, in Hebrew, *mishkan*, a word deriving from the same verbal root as *Shekinah*.[2] Abraham's altars are habitacles of *Shekinah* and they represent the *principial* point's "modes," "the vital centers" reflecting the supreme center, the sacred knots of the texture that keep the holy land holy. If the profane point of view starts to prevail, *Shekinah* will withdraw and the holy land will become an empty shell. To maintain *Shekinah* in place, while the cycle is declining, there is a need for more elaborate rites, performed more often, an increase of spiritual degrees to allow the spiritual influences to operate efficiently and guard the land holy; however, in the end, due to the lack of righteous men, the center will disappear underground and the holy land will become a barren desert, a land of desolation.

With the descent of the cycle, the spiritual influences and the spiritual center become "lost," hidden and inaccessible, which is the same as saying that they withdraw underground, into the depths of the earth, in the heart of the mountain. The various traditions allude to a mysterious subterranean world, which invisibly communicates with all the regions of the earth; the cavern is the cave of the heart, the center of the being, the Hermetic *athanor*, the initiatory place, *Beith-El* (the house of God), but also Agarttha, previously called Luz, the mysterious city situated underground, that is, totally hidden.[3]

---

[1] Guénon, *Symboles fondamentaux*, pp. 110-112.
[2] Vulliaud, I, p. 493, Guénon, *Le Roi*, p. 25.
[3] The symbol for the mountain is a triangle whose apex is headed up, the cave is a triangle whose apex is headed down (Guénon, *Symboles fondamentaux*, p. 224). Guénon adds: "If we want to represent the cave as located inside (or in the heart, one might say) of the mountain, we just have to carry the inverted

"We should (...) talk about something that is hidden rather than truly lost, since it is not lost for all and some people still possess it in full; and, if so, others will always have the possibility to regain it, provided they seek it in an appropriate manner, that is, if their intention is directed in such a way that, by harmonic vibrations[1] that this intention awakens in accord with the law of 'concordant actions and reactions,' it can bring them into effective communication with the spiritual supreme center. The direction of this intention, in all traditional forms, has its symbolic representation, and we refer here to the ritual orientation: the latter, indeed, is the proper direction to a spiritual center, which, whatever this it may be, is always a true image of the Center of the World."[2]

---

triangle within the straight triangle, such so that their centers coincide. (...) On the other hand, if one makes the sides of the inverted triangle equal to half those of the right triangle, the small triangle divides the surface of the large in four equal parts" (pp. 225-226): it is a central and luminous image of the *Royal Arch*.

[1] In the next chapter we will show the symbolic importance of the harmonic vibrations with regard to the center and the spiritual influences.

[2] Guénon, *Le Roi*, p. 69.

II

THE CENTER AND THE SPIRITUAL INFLUENCES

A true spiritual center should be regarded as representing the Will of God in this world[1]; for the Far-Eastern tradition this center is the "Invariable Middle" where the "Activity of Heaven" manifests itself[2]; for the Jewish tradition, "such a center, established according to regularly defined conditions, should be the place of divine manifestation," that is, of *Shekinah*.[3]

Obviously, the Center of the World, as a contact point between Heaven and Earth, is the custodian of the "non-human" elements that constitute the immortal essence of the symbols, rituals, of the traditional doctrine itself (generating the symbolic form of its expression[4]), and of the man who transmits the doctrine or performs a rite or follows an initiatory path.[5] Of course, all this is much more complex and we cannot deal but with the essential aspects, following the example of the Daoist painter for whom the void rather than the drawing is essential. The center transmits the "non-human" elements and maintains them using a continuous vibration in harmony with the vibrations of the world; yet, when decadence becomes preponderant, when the revolt against spiritual authority finds

---

[1] René Guénon, *Aperçus sur l'initiation*, Éd. Trad., Paris, 1992, p. 68.
[2] Guénon, *Symboles fondamentaux*, p. 88.
[3] Guénon, *Le Roi*, p. 23.
[4] Guénon, *Aperçus sur l'initiation*, p. 285.
[5] "The intervention of a 'non-human' element can define, in a general way, everything that is authentically traditional" (*ibid.* p. 26).

success, and the profane point of view becomes dominant, this transmission ceases and the vibrations withdraw into the center, the solidified world being unable to respond to them, and so, gradually, symbols are no longer understood, the rituals are altered or forgotten, but mainly they lose their effectiveness, and man, for his part, is captured by the "ego" with such a force that he denies any "non-human" reality.

The "non-human" elements descend along the supreme Way (*Dao* or the Axis of the World), yet without leaving the absolute Center (identical to the Principle itself). The Way, represented by the vertical axis, Guénon said, refers to the "Universal Man," which is identical with the Self, while the "Truth," represented by one of the two horizontal axes of the spatial cross, is connected to the "intellectual man," and the "Life" (represented by the other horizontal axis) to the "corporeal man."[1] What we have here is René Guénon's metaphysical interpretation of Saint John's Gospel, in which *Logos* or the Will of Heaven at work is "the Way, the Truth and the Life."[2] At the beginning of Saint John's Gospel we find the well-known words: "In the beginning was the Word, and the Word was with God, and the Word was God. In him was life, and the life was the light of men. This was the true light, which lighteth every man that cometh into the world." Of course, light and life are foremost symbols replacing the *principial* Light and the *principial* Life, a symbolism that was outlined by René Guénon in an article entitled *Verbum, Lux et Vita*.

---

[1] Guénon, *Le symbolisme de la croix*, p. 123. The intellectual man together with the corporeal man constitutes the "veritable man."

[2] "Jesus saith unto him, I am the way, the truth, and the life" (*John* 14:6). It is very important to note that it is in Saint John's Gospel that we find the most essential symbolic attributes of Christ: "I am the living bread which came down from heaven" (6:51); "Whoever believes in me, streams of living water will flow from his womb" (7:38); "I am the light of the world; whoever follows me will not walk in darkness, but it will have the light of life" (8:12); "While I am in the world, I am the light of the World" (9:5); "I am the door: by me if any man enter in, he shall be saved" (10:9); "I am the good shepherd" (10:11); "I am the way, the truth and the life" (14:6); "I am the true vine, and my Father is the vinedresser" (15:1). We should note the importance of light and life.

We have already seen that God through His Word becomes the World's Center, and *Manu*, who is the Cosmic Intelligence, reflects the spiritual Light.[1] In addition, Guénon said that the spiritual center is the place of divine manifestation, always represented as Light. This light is *Shekinah*, the "real presence" of Divinity, and "it should be noted that the passages from Scripture where [*Shekinah*] is specifically mentioned are mainly those concerning the foundation of a spiritual center: the construction of the Tabernacle, the construction of Solomon's and Zerubbabel's Temple."[2]

Shekinah is the immediate presence of God into the world and man; it is the universal light produced in the first day.[3] "The mystery of her function is written *Yod Beth Koph*, that is, by using the initials of *Ykuda* (Unity), *Beraka* (Benediction) and *Kedoscha* (Holiness). Shekinah is the mediator by which the Union between the inferior world and the Holy One, blessed be He (*Kaddosh-Baruch-Hu*), is realized, she is the organ of the blessings from top to bottom and from bottom to top, and finally she is the principle of Sanctification. The blessings are transmitted through Shekinah along the arteries of the universal body."[4]

This text from Vulliaud is very important. René Guénon, in the chapter *La « Shekinah » et « Metatron »* of his *Le Roi du Monde*, after introducing the two heavenly mediators, said: "We don't want to elaborate here upon the theory of 'spiritual influences' (we prefer this expression, instead of the word 'blessings,' to translate the Hebrew *berakoth*, especially since this is precisely the very sense of the Arabic word *barakah*)."[5] Therefore, we can rephrase the last sentence of Vulliaud's text and state that the spiritual influences are transmitted to the spiritual center through *Shekinah*. The spiritual influence is the "non-human" element carried by rites, their completion involving "the action

---

[1] Here we find another essential and "central" element: the Intelligence.
[2] Guénon, *Le Roi*, p. 23. Vulliaud, *La Kabbale Juive*, I, p. 489.
[3] Vulliaud, *La Kabbale Juive*, I, p. 501.
[4] Vulliaud, *La Kabbale Juive*, I, p. 508.
[5] Guénon, Le Roi, p. 23.

of an influence belonging to a superior order, which can be properly called 'non-human,' both in the case of the initiatory rites and of the religious ones."[1]

"We therefore say this: any religion, in the true sense of that word, has a 'non-human' origin and is organized in such a way that allows it to be the depository of an element also 'non-human,' received from this origin; this element, which belongs to the order of what we call spiritual influences, exerts its effective action through appropriate rites, and the performance of these rites in order to be valid, i.e. to provide a real support to the influence in question, requires a direct and uninterrupted transmission. (…) If this is true for the exoteric order, more so will it be for a higher order, that is, for the esoteric order."[2] Guénon added: "The rite always contains a teaching in itself, and the doctrine, because of its 'non-human' character, also carries a spiritual influence within"[3]; "It should be added that, when it comes to really traditional rituals and symbols, their origin is similarly 'non-human.'"[4]

The interruption of the spiritual influence's transmission is equivalent to a fissure or a break in the chain of the initiatory process (in Arabic *silsilah*, in Hebrew *shelsheletk*) comparable to the fissures produced by the malevolent influences in the "Great Wall"[5]; "The wicked actions form what Kabbalah calls a a *breach* in the Holy name of God. (…) Thus, the blessings cannot longer flow into the Sephirothic channels, being blocked by sin. Since the flow cannot longer take place, the 'Holiness' cedes his empire to the Demon."[6] In fact, there is, as always, a dual operation, heavenly and earthly: the breach in the transmission of the spiritual influence is doubled by the withdrawal of this influence, by its reabsorption into the center, which then becomes an occulted center because of the world's

---

[1] Guénon, *Aperçus sur l'initiation*, p. 54.
[2] Guénon, *Aperçus sur l'initiation*, pp. 55-56.
[3] Guénon, *Aperçus sur l'initiation*, p. 285.
[4] Guénon, *Aperçus sur l'initiation*, p. 115.
[5] Guénon, *Le règne*, p. 230.
[6] Vulliaud, *La Kabbale Juive*, I, p. 508.

desecration. "Shekinah goes away or comes near the man and the universe, according to the degree of individual and collective Holiness or purity, which is one of the characteristics of Holiness."[1]

The first righteous man [*tsaddiq*] who brought *Shekinah* down here was Abraham.[2] "Shekinah is benediction, blessed and again blessed; and she blesses. Abraham and his sons have received her in heritage."[3] We understand now Abraham's crucial role as builder of altars: only a blessed and consecrated man, blessed with *Shekinah* and filled with spiritual influence, has the requisite qualifications to build an altar, that is, a "House of God," and to perform the rites that awaken the spiritual influence's operation in order to dedicate the altar.

In the Islamic tradition, Abraham appears as a *hanîf*, a pure[4] and unified sage, dedicated to the absolute and unlimited adoration of the metaphysical Truth.[5] "Who has a better religion than he who resigns his face to God, and does good, and follows the Rule of Abraham, as a 'Hanîf'? For God took Abraham as a bosom friend (*khalîl*)."[6] As God's "bosom friend," Abraham is the beneficiary of spiritual realization implying the supreme degree of love[7] and he is, after Gilis, "a figure of the universal priest."[8] Abraham is called "a bosom friend," Ibn 'Arabî teaches us, because he "penetrated" and assimilated the Qualities of the divine Essence; or, his name implies that God has essentially penetrated Abraham's form.[9] For Gilis, when Abraham is regarded as penetrating and

---

[1] Vulliaud, *La Kabbale Juive*, I, p. 509.
[2] Vulliaud, *La Kabbale Juive*, I, p. 509.
[3] Vulliaud, *La Kabbale Juive*, I, p. 510.
[4] "In truth, Ibrahim was neither Jew nor Christian, but it was a *hanîf*" (*Qur'an* 3:67).
[5] Michel Vâlsan, *L'Islam et la fonction de René Guénon*, Les Éditions de l'Œuvre, 1984, p. 131. "Abraham was one. He was so called because no one else of his contemporaries attained the virtue of faith in God" (*Zohar*, I, 85 b).
[6] *Qur'an* 4:124.
[7] Out of love for God Abraham agrees to sacrifice his son.
[8] Arabî, *Le livre des chatons*, I, p. 177.
[9] Muhyi-d-dîn Ibn 'Arabî, *La sagesse des prophètes*, Albin Michel, Paris, 1974, tr. Titus Burckhardt, p. 75, Arabî, *Le livre des chatons*, I, tr. Gilis, p. 168.

encompassing all the qualities of the divine Essence, it means that he realized in an initiatory manner the name of God – it is the ascendant spiritual realization; when we consider that it is God who penetrates the actual reality of Abraham's form, it signifies the descendant realization.[1]

This twofold initiatory process could also be found in the Jewish tradition. Abraham is the beloved of God and he loves God because he loves justice (*tsedeq*). Abraham is the righteous one (*tsaddiq*), unlike a lot of people who are far from God and refuse to approach God and therefore they are far from justice; and because they are far from justice, they are far from peace and they are not at peace.[2] What is suggested here is Abraham's Abraham's special quality related to Justice and Peace, that is, to the Center of the World.[3] The extraordinary relationship between Abraham and God and his ability to bring *Shekinah* down here are understandable if we fathom that he is the "righteous" one and loves justice as much as *Melki-Tsedeq*, his superior; however, it is unclear why others are far from justice, but not Abraham. Abraham's spiritual realization follows the three degrees of the soul[4]: "there is a throne resting on a throne, throne, and a throne for the highest,"[5] that is, *nephesh*, the lowest lowest stirring, to which the body is attached, *ruah* and *neshamah*, but it is also a twofold initiatory process. "When Abram entered the land God appeared to him and he received there a *nephesh*, and built an altar to the corresponding grade of divinity. Then 'he journeyed to the South,' receiving a *ruah*. Finally he rose to the height of cleaving to God through the medium of the *neshamah*, whereupon he 'built an altar to the Lord.' (...) He went

---

[1] Arabî, *Le livre des chatons*, I, p. 179.
[2] *Zohar*, I, 76 b. "*Melki-Tsedeq* is thus both king and priest. His name means 'king of Justice,' and he is also king of *Salem*, that is of 'Peace'; we find here, before all, 'Justice' and 'Peace,' that is, precisely the two fundamental attributes of the 'Lord of the World'" (Guénon, *Le Roi*, p. 49); and René Guénon added that Salem is the equivalent term for Agarttha. Let us note that *Melki-Tsedeq* has blessed Abraham.
[3] The Garden of Eden is the residence of the righteous (*Zohar*, I, 77 b).
[4] "Scrutinizing these degrees of the soul, you can enter the supreme wisdom."
[5] *Zohar*, I, 83 b.

down to Egypt (...) he 'went up' from Egypt literally, strengthened and confirmed in faith, and reached the highest grade of faith. Thenceforth Abram was acquainted with the higher Wisdom."[1]

There is no doubt that Abraham, the "righteous,"[2] was an initiate. He benefited from the contribution of the spiritual influences with an upward progression, but he also had the ability or a natural inclination (the initiatory "qualification")[3] corresponding to the degree of knowledge inherited from his previous states of existence. Ibn 'Arabî, in the chapter on Abraham explained: "It is not within the possibilities of each living creature in this world for God to open the eye of his (intuitive) intelligence to see the reality as it is; there are those who know and others who are unaware. So God did not want to lead them all and He did not guide them all, and He does not wish to do so."[4]

We have previously seen that a secondary center must be prepared becoming empty of all the obstacles and of all the "bad" influences and must be purified in an efficient way, in order to become the house of God. In a similar manner the soul must be prepared to receive the presence of God (as Maître Eckhart said); Abraham is a pure sage not only in the sense of perfection, but also as a purified man who has been prepared to receive *Shekinah*, and Guénon has shown that the initiatory trials are essentially rites of purification, preliminary or preparatory rites to the initiatory process itself.[5] "*Melki-Tsedeq* is represented represented as superior to Abraham, since he is the one who gives the blessing. (...) The 'benediction' of which he speaks is actually the communication of a 'spiritual influence,' in which Abraham would henceforth participate; and it can be noted that

---

[1] *Zohar*, I, 83 b.
[2] "And Abraham said, Oh let not the Lord be angry, and I will speak yet but this once: Peradventure ten righteous shall be found there. And he said, I will not destroy it for ten righteous' sake" (*Genesis* 18:32).
[3] Guénon, *Aperçus sur l'initiation*, p. 29.
[4] Arabî, *La sagesse des prophètes*, p. 79.
[5] Guénon, *Aperçus sur l'initiation*, p. 175.

the formula used placed Abraham in a position of direct communication with 'God Most High.'"[1] The immediate relationship between Abraham and God is a direct communication with the supreme Center, intellectual, intuitive, permanent and non-religious in nature, important specifications because "There are many believers and few intuitive apprehenders."[2] René Guénon said: "Let us accept that (...) there is a genuine communication with the superior [supra-individual or super-individual] states; it is still far from sufficient to characterize the initiatory process. Such a communication exists also in the purely exoteric rites, especially in the religious rites; it should not be forgotten that in this case also, spiritual influences, and not merely psychological, really come into play, although for different purposes than those related to the domain of the initiatory process. (...) With regard to the initiatory process, the simple communication with the superior states cannot be regarded as an end but only as a starting point; if this communication must be established first by the action of a spiritual influence, it is to allow next the taking in possession of these states, and not simply, as in the religious order, to make descend a 'grace' that connects us to them in a certain way, but without allowing us to penetrate them."[3]

Abraham is an "intuitive believer" and not simply a believer, which requires a complete spiritual realization. The man, as a man, can reach the Terrestrial Paradise, the center of the integral human state, regarded with the total extension of its possibilities, but to reach Heavenly Paradise, the supreme Center, someone must realize the supra-human states, and "in the symbolism of the cross, the first of these two realizations is represented by the development of the undefined horizontal line (the Truth and the Life), and the second by the vertical line (the Way)."[4] Direct communication with the superior states,

---

[1] Guénon, *Le Roi*, p. 50.
[2] 'Arabî, *La sagesse des prophètes*, p. 80.
[3] Guénon, *Aperçus sur l'initiation*, pp. 26-7.
[4] René Guénon, *Autorité spirituelle et pouvoir temporel*, Véga, Paris, 1976, pp. 98, 102.

Guénon said, is based on the existence of transcendent faculties with respect to the individual, regardless of the name given to them, "intellectual intuition" or "inspiration."[1] Metaphysical truths, Guénon stressed, cannot anymore be fathomed by an individual faculty, and the immediate nature of this operation is what makes it an intuitive faculty, called either intellectual intuition or pure intellect[2]; metaphysics affirms the fundamental identity of *knowing* (French "connaître") and *being* (French "être"), Guénon added, and this identity is essentially inherent to the very nature of the intellectual intuition, that is, this identity is realized exclusively by intellectual intuition.[3]

To understand how the spiritual influences, descending from the Center, transform the human being and contribute to the realization of the identity *knowing = being*, which, in essence, means taking actual possession of the supra-individual states and much more, it was necessary to remember these few remarks of René Guénon, in order to emphasize the importance of the pure intellect and intellectual intuition.

The first degree of *Âtmâ*'s manifestation, the Hindu tradition affirms, is the pure intellect (*Buddhi*), *Prakriti*'s first production, a principle of universal order, that is, a principle that exceeds any individual state, including the human condition.[4] Even if *Buddhi* is "the greatest" (*Mahat*) of *Prakriti*'s productions, it must be understood that there is, similarly to the generation of integral universal manifestation, a simultaneous operation of *Purusha* (metaphysical order) and *Prakriti* (cosmological order). "The origin and development of manifestation can be envisaged at one and the same time in an ascending and a descending

---

[1] "The transcendent intellect, to comprehend directly the universal principles, must be itself universal; it is no longer an individual faculty" (René Guénon, *La métaphysique orientale*, Éditions Traditionnelles, Paris, 1993, p. 11).

[2] René Guénon, *Introduction générale à l'étude des doctrines hindoues*, Guy Trédaniel, 1987, p. 94. Without this pure intellectual intuition there is no real metaphysics (Guénon, *La métaphysique orientale*, p. 11). The intellectual intuition corresponds to *Buddhi* (René Guénon, *L'homme et son devenir selon le Védanta*, Éditions Traditionnelles, Paris, 1991, p. 119).

[3] Guénon, *Introduction générale*, pp. 144-145.

[4] Guénon, *L'homme et son devenir*, p. 71.

sense. If this is so, it is because manifestation does not proceed only from *Prakriti*, from which its entire development is a gradual transition from potency to act and can be described as an ascending process; but proceeds, in reality, from the two complementary poles of Being, that is, from *Purusha* and *Prakriti*."[1]

For this, if we consider the "Self" (*Âtmâ*) as the spiritual Sun that shines in the center of the total being – where the center is identical to *Âtmâ* – Buddhi will be the ray directly originated from this sun, illuminating the individual human state in its entirety.[2] In 1947, René Guénon wrote again about this subject, in his article *Spirit and Intellect*, where he identified *Buddhi* with *Spiritus* and stressed that the intellect is never an individual mind, in the same way as the spirit[3] is never really "embodied," and there is really no difference between the spirit and the intellect[4]; in addition, he clarified the nature of *Buddhi*: "from a *principial* viewpoint, *Buddhi* would appear as the beam of light originated directly from the spiritual Sun, which is *Âtmâ* itself; so one can say that it is also the first manifestation of *Âtmâ*."[5]

Buddhi is the celestial or solar Ray linking the centers of all the individual and supra-individual states to the absolute Center, including the Center of the World or the Earthly Paradise[6]; accordingly, the center of the integral human being is connected through *Buddhi* to "all manifested states, individual and non-individual, of the same being, and, beyond them, to the center itself."[7] Therefore, the spiritual realization implies, in order to achieve the supreme goal, a unique path that is this ray,

---

[1] René Guénon, *Initiation et réalisation spirituelle*, Éd. Traditionnelles, Paris, 1980, pp. 241-2, Mircea A. Tamas, *About the Yi Jing*, Rose-Cross Books, Toronto, pp. 107-8.
[2] Guénon, *L'homme et son devenir*, p. 72.
[3] Here "spirit" must be understood as *Spiritus*, intimately related to "spiritual."
[4] Thus, the spiritual influence is strongly related to the intellect.
[5] René Guénon, *Mélanges*, Gallimard, Paris, 1976, p. 34.
[6] The realization of the integral individuality is represented as retaking possession of the Earthly Paradise and a restoration of the "primordial state" (Guénon, *La métaphysique orientale*, p. 17).
[7] Guénon, *L'homme et son devenir*, p. 72.

Buddhi, through which the being is connected to the spiritual Sun; and Guénon stressed that "regardless of the apparent diversity of existing ways at the point of departure, they must all come together sooner or later into this unique 'axial' way."[1]

It is extremely important to note here an essential point: when we speak of a total spiritual realization and of the access to the absolute Center through the "axial" way, it is no longer the human being that we have to consider, but the total being.

Reason or the mind is a human faculty.[2] When a man is about to die, the Hindu tradition teaches us, the speech, followed by the rest of the ten external faculties, is absorbed in the internal sense (*manas*, the mind); this one then withdraws into the "vital breath" (*prâna*), together with all the vital functions that are inseparable of life itself; the "vital breath" withdraws after that into the "living soul" (*jîvâtmâ*), which is the reflection of *Âtmâ* in the individual human being, mitigated by *Buddhi*.[3] Yet, it is not as man or *jîvâtmâ* that the human being can reach the supreme Center and metaphysical knowledge; it is because "this being, which is human in one of his states, at the same time is something else and more than a human being[4]; and and this effective awareness of the supra-individual states is the real object of metaphysics, or, better yet, is metaphysical knowledge itself."[5] So, when we consider light and life (or the "vitality") as features describing the spiritual influences, it must be clear that this is just a symbolic meaning that we extend beyond the human being, even considered in all extension of its possibilities. It should be noted that light and "vital breath" belong to the subtle manifestation, which in the Hindu tradition is called *Taijasa*, the "Luminous."[6]

---

[1] Guénon, *Mélanges*, p. 36.
[2] Guénon, *La métaphysique orientale*, p. 11.
[3] Guénon, *L'homme et son devenir*, pp. 60, 73, 145-6. But the intellect is not different from *Âtmâ* (Guénon, *L'homme et son devenir*, p. 121).
[4] "But Jesus beheld them, and said unto them, With men this is impossible; but with God all things are possible" (*Matthew* 19:26).
[5] Guénon, *La métaphysique orientale*, p. 11.
[6] Guénon, *L'homme et son devenir*, pp. 107, 109.

Âtmâ "is presented in two manners: on the one hand the Plenitude-of-Vital-Breath (*Prâna*), on the other hand the Sun. They are also the two Ways"[1]; "The Vital-Breath is *Brahma*."[2] *Prâna* is a substitute for *Brahma*, as *Hiranyagarbha* is one for the Principle, and as the Ether (*akâsha*) is a substitute for *Âtmâ*,[3] but all this is related to the initiatory teaching method that leads the neophyte step by step to *Brahma* (*Arundhati-darshananyâya*). *Âtmâ*, as René Guénon explained, is indeed the universal Center containing all things, but, by reflecting itself in the human manifestation, it appears to be "localized" in the center of individuality, and even, more precisely, in the center of its bodily modality, although this reflection is only an appearance.[4] For this reason, it is said that *Brahma* is the one that resides in the vital center of the human being, which is not only the center of corporeal individuality, but also that of integral individuality (capable of indefinite extension in its domain), a center identical to the heart, which is not only the center of life, but the symbolical seat of universal Intelligence or Intellect.[5] "Thus, the the one that resides in the vital center, from a physical point of view, is the Ether; from a psychical viewpoint, it is the 'living soul'; (…) from the metaphysical point of view, it is the *principial* and unconditioned 'Self.' Therefore it is indeed the 'Universal Spirit' (*Âtmâ*), which is in reality *Brahma* itself."[6]

From a *principial* perspective, the formal or subtle manifestation is contained in *Hiranyagarbha* – the "Golden Embryo," which is *Brahmâ* as determination or effect (*kârya*) of

---

[1] *La Maitrâyanîya Upanishad*, tr. Jean Varenne, I, B. The two ways are: *pitri-yâna* that leads to Moon (corresponding to *Prâna*) and *dêva-yâna* (corresponding to Light). "All those who leave this world arrive to the Moon. Due to their vital breaths the moon grows" (*Kaushitaki Upanishad*, I. 2).
[2] *Kaushitaki Upanishad*, II. 1.
[3] *Chândogya Upanishad*, 8. 1. 1.
[4] Guénon, *Initiation et réalisation spirituelle*, p. 237.
[5] Guénon, *L'homme et son devenir*, pp. 42-3.
[6] Guénon, *L'homme et son devenir*, p. 46.

Brahma enveloped in the "World's Egg"[1]; *Hiranyagarbha* is "the synthetic assembly of life" (*jîva-ghana*), and Guénon, based on the connection of the subtle state with life, called it "Universal Life," an appellation akin to the saying inscribed in the *Gospel of Saint John*, "And Life was the Light of men."[2] The "Universal Spirit" (*Âtmâ*) projects the "Celestial Ray," which is reflected upon the mirror of the "Waters" and gives birth to *Hiranyagarbha*, the determination of the "Non-Supreme" *Brahma*.[3]

René Guénon explained: "The symbolic journey, accomplished by the being in its process of gradual liberation, from the termination of the coronal artery (*sushumnâ*), constantly communicating with a ray of the spiritual Sun, to its final destination, is following the Way that is marked by the route of this ray traveled in the opposite direction (following its reflected direction) to its source (the Center)."[4]

It is the "divine journey" (*dêva-yâna*), but Guénon is really very cautious, knowing the confusion made between "salvation" and "liberation"; thus he insists that this divine voyage, "which refers to the effective identification of the (integral) individuality's center with the very center of the total being, residence of the Universal *Brahma*,"[5] is valid only if such an identification was not realized during the earthly life (*jîvan-mukti*), nor even at the time of death (*vidêha-mukti*), so that the being "may remain in the cosmic order and not reach the actual possession of the transcendental states, which in fact

---

[1] Guénon, *L'homme et son devenir*, p. 112. Guénon underlined that *Hiranyagarbha* "has a very close meaning to that of *Taijasa*, because the gold, according to the Hindu doctrine, is the 'mineral light.'"

[2] We saw that the light and life are quoted repeatedly in this Gospel. Life, René Guénon said, "even if regarded in all extension which it is capable, is nothing more than one of the special conditions of the state of existence to which belongs the human individuality; the domain of life does not, therefore, exceed the possibilities that constitute this state" (Guénon, *L'homme et son devenir*, p. 113).

[3] Guénon, *Le symbolisme de la croix*, p. 127.

[4] Guénon, *L'homme et son devenir*, p. 167.

[5] Guénon, *L'homme et son devenir*, p. 169.

The Center and the spiritual influences 43

represents the veritable metaphysical realization."[1] In this case, *Brahma-Loka*, the ultimate goal of the divine journey, is not the center of *Brahma*, but *Hiranyagarbha's*, the principle of the subtle manifestation, that is, of the whole domain of human existence in its entirety, a "cosmic center" virtually identified with the center of all worlds[2]; the "salvation," Guénon stressed, is actually attaining *Brahma-Loka* as residence of *Hiranyagarbha*.[3]

Four years later, René Guénon added: "The 'Heavenly Paradise' is essentially *Brahma-Loka*, identified with the 'spiritual Sun' and the 'Earthly Paradise' is described as touching the 'sphere of the Moon.'"[4] Therefore, to really attain the "Heavenly Paradise" or the supreme Center, the being must travel beyond *Hiranyagarbha*, beyond the "current of forms" (or the world of forms),[5] beyond the light and beyond the life: it is the "axial" or "vertical" journey following the heavenly Ray; it is the "Middle Way" or the "Way of Heaven," symbolized by the vertical axis envisaged in the ascendant direction,[6] aiming at the identification of the center of human state with the center of the total being, when the earthly pole differs no more from the heavenly pole. "Since then, there is, strictly speaking, no more axis, as if this being, while he gradually identified himself with the axis, somehow had 'absorbed' it reducing it to a unique point" – the Center.[7]

"And now, the channel *anabhu* [the vertical axis]: it is the one that leads to the Sun [Center] the sacrifice which is offered in the fire. The juice that flows comes back in the form of rain [the spiritual influence]: it is the *udgitha*. Thanks to it there are the vital breaths and thanks to the breaths the creatures. In this context it was said: 'I am offering this sacrifice in the fire, it will

---

[1] Guénon, *L'homme et son devenir*, p. 175.
[2] Guénon, *L'homme et son devenir*, pp. 177-8.
[3] Guénon, *L'homme et son devenir*, p. 189.
[4] Guénon, *Autorité*, p. 102.
[5] Guénon, *La métaphysique orientale*, p. 19.
[6] "The vertical axis is the metaphysical locus of the manifestation of the 'Heaven's Will' and it crosses every horizontal plane in its center" (Guénon, *Le symbolisme de la croix*, p. 120).
[7] Guénon, *La Grande Triade*, p. 209.

reach the Sun; and the Sun will rain upon me by its rays [celestial ray], and I will have food; and from it the creatures will be born!"[1] This food is the metaphysical Knowledge assimilated by intellectual intuition and revealed down here by the spiritual influences descended in the form of solar rays and rain. "But he answered and said, It is written, Man shall not live by bread alone, but by every word that proceedeth out of the mouth of God."[2]

Buddhi, the celestial or solar Ray, the vertical axis, is the Way leading the initiate to the Center, but it is the same vertical Way where the Logos or the Will of Heaven is manifested, and it is also the Way through which the spiritual influences come down into the Three Worlds and into all the degrees of the universal manifestation. "Shekinah is holy, sanctified, sanctifying. Becoming holy body, soul and spirit, it means to become the temple of Shekinah."[3] *Shekinah* operates simultaneously in the Three Worlds, that is, the spiritual influences operate at the same time in the informal or supra-individual manifestation, in the subtle manifestation and in the corporeal manifestation, or in *Spiritus*, *Anima* and *Corpus*. The celestial Ray crosses all the states of the being, marking the midpoint of each by its trace on the horizontal plane, but according to Guénon, "this action of the celestial Ray is effective only if it produces, by its reflection upon one of these planes, a vibration, which propagated and amplified in the being in its entirety, illuminates its chaos, human or cosmic."[4] In a similar way, the spiritual influences come down with the celestial Ray and are propagated from the

---

[1] *La Maitrâyanîya Upanishad*, III, P. "It being necessary for the impulse from below to set in motion the power above. So vapour first ascends from the earth to form the cloud. Similarly, the smoke of the sacrifice rises and creates harmony above, so that all unites, and in this way there is completion in the supernal realm. The impulse commences from below, and from this all is perfected" (*Zohar*, I, 35 a). "Verily I say unto you, Whatsoever ye shall bind on earth shall be bound in heaven: and whatsoever ye shall loose on earth shall be loosed in heaven" (*Matthew* 18:18).

[2] *Matthew* 4:4.

[3] Vulliaud, *La Kabbale Juive*, I, p. 510.

[4] Guénon, *Le symbolisme de la croix*, p. 126.

absolute Center into the center of each world and subsequently into the secondary spiritual centers following an uninterrupted flow,[1] and the most intelligible way to represent the spiritual influence is a vibratory motion.

René Guénon symbolized the Universal Possibility (which is not different from Infinity[2]) by the "universal spherical vortex," identical to a vibratory movement: "The deployment of this spheroid is, in fact, nothing more than the indefinite propagation of a vibratory movement, not only in a horizontal plane, but throughout the three-dimensional extension, where the starting point of this movement could be viewed as being the center."[3] And Guénon finished with an essential clarifycation: "If we consider this extension a geometrical symbol, that is, a spatial one, of the total Possibility, we should obtain the representation (insofar this illustration is possible) of the universal spherical vortex, along which flows the realization of all things, and which the metaphysical tradition of the Far-East calls Dao, that is, the 'Way.'"[4]

It should be clearly understood that the celestial Ray has no specific direction; it is the "seventh ray" – the Way through which the being, after completing the cycle of manifestation, returns to non-manifestation and is effectively united with the Principle,[5] the same Way through which the spiritual influences are spread downwards; this Way is the Center, and it is also the universal spherical vortex, represented by an absolute vibratory movement. Guénon compares the spiritual influence's vibration with the *Fiat Lux* that illuminates the chaos, an illumination that "is precisely constituted by the transmission of the spiritual

---

[1] It is also the Hesychastic initiatory way, where the uninterrupted Prayer of Heart compels the intelligence, that is, the spiritual influence, to descend into the heart: "Lord [the supreme Center], Jesus Christ [the celestial Ray], Son of God [the Center of the World], have mercy [the spiritual influence] on me, the sinner [human being]."
[2] René Guénon, *Les états multiples de l'être*, Guy Trédaniel, Paris, 1984, p. 18.
[3] Guénon, *Le symbolisme de la croix*, p. 112.
[4] Guénon, *Le symbolisme de la croix*, p. 112.
[5] Guénon, *Symboles fondamentaux*, p. 350.

influence,"[1] and the initiatory trials, performing the "purification," bring the being back "to a state of undifferentiated simplicity, comparable to that of the *materia prima*, in order to be able to receive the vibration of the initiatory *Fiat Lux*. The spiritual influence, when transmitted, will give him this first 'enlightenment,' and must not encounter in him any obstacle due to the inharmonious 'performance' of the profane world."[2] We might add that this is the profound reason for fasting or for practicing *Hatha-Yoga*[3]: it is necessary to purify *Corpus* and *Anima* to be able to absorb the vibration of the spiritual influence.

The idea of a vibratory movement has already been used by Matgioi who, speaking of "concordant actions and reactions," described them as an "undulatory vibration,"[4] but it was Nicholas of Cusa who introduced the spirit of God's gradually descending movement, from universal to particular.[5] René Guénon connected the vibratory movement with the production of *bhûtas*, considering the five elements as different vibratory modalities, the elementary movement from which the ether gives birth to the other elements being the prototype of the vibratory movement.[6] We could consider a similar vibratory vibratory movement to describe the spiritual influences that descend by countless degrees and which, arriving at the state of the human being, will operate an essential transformation of the

---

[1] Guénon, *Aperçus sur l'initiation*, p. 34, *Le symbolisme de la croix*, p. 141. The transmission of a spiritual influence has to take place following specific laws, rigorously defined, which present a certain analogy with the laws that govern the forces of the corporeal world, and it is this analogy "which has allowed us, for example, to speak of 'vibration' regarding *Fiat Lux* through which the chaos of the spiritual potentialities is illuminated and organized" (Guénon, *Aperçus sur l'initiation*, pp. 35-6).

[2] Guénon, *Aperçus sur l'initiation*, pp. 176-7. According to Meister Eckhart, the expulsion of the merchants from the temple is the symbolic representation of this purification.

[3] René Guénon, *Études sur l'hindouisme*, Éditions Traditionnelles, Paris, 1979, p. 141.

[4] Matgioi, *La voie rationnelle*, Éditions Traditionnelles, Paris, 1984, p. 144.

[5] Cusa, *De la docte Ignorance*, p. 148.

[6] Guénon, *Mélanges*, p. 113, *Études sur l'hindouisme*, p. 64.

corporeal and subtle elements, with the condition that they are purified and ready to receive the initiatory vibration, the transformation implying that all these elements become valid supports for the spiritual influences.[1] Thus, vital breath, flesh and blood,[2] movement and gestures, the rhythmical words, and so on, are not only symbols in the initiatory or even religious rites, but also effective supports for the spiritual influences and, what is even more important, they really are elements transformed by the "non-human" vibration and prepared for the purely intellectual "great voyage."

"As Shekinah resides in man, says the Zohar,[3] many sacred armies surround him. The angels [the spiritual influences][4] are degrees of the divine presence.[5] Whoever possesses the presence of God in the highest degree is the Holy Man [the Universal Man], Shekinah being the totality of the degrees."[6] There is a hierarchy of the spiritual influences as well as there is an angelical hierarchy,[7] which implies that, for a total spiritual realization, for obtaining the unconditioned absolute state – the Absolute Center – designated by the Hindu doctrine as "Liberation," it is necessary to surpass all these individual components and climb up along the angelical hierarchy beyond these supports and adjutants, into the domain of pure intellect where *Shekinah* is the totality of the degrees. For example, a *mantra* has no effect if it has not been received directly from the mouth of an authorized *guru*, because it is not "enlivened" or "animated" by the presence of the spiritual influence for which

---

[1] "The spiritual influences themselves, to start operating in our world, must necessarily take suitable 'supports,' first in the psychical order and then in the corporeal order itself" (Guénon, *Le règne*, p. 248).

[2] The blood, for example, is the support of the vital energy, and the vital energy is the support of the spiritual influence.

[3] I, a 166.

[4] Guénon said: "The 'angelic' states are in fact identical to the supra-individual individual states of the being" (*Autorité*, p. 100). "Are they [the angels] not all ministering spirits, sent forth to minister for them who shall be heirs of salvation?" (*Hebrews* 1:14).

[5] *Zohar*, III, 52 b.

[6] Vulliaud, *La Kabbale Juive*, I, p. 510.

[7] There is also a hierarchy of spiritual centers.

it is only intended to be the vehicle; it is, Guénon added, the communication of something "vital," related to the vital breath (*prâna*) and to the voice, but *prâna* itself is just the vehicle or the subtle support for the spiritual influence that is transmitted from *guru* to the disciple[1]; also, the expression of an idea in a vital mode is after all a symbol like any other.[2] The means of the metaphysical realization must be within the reach of man; "so the forms belonging to this world where his present manifestation is situated will be the support for the human being to rise above this world itself"[3]; "The blessing cannot descend from above to an empty place, but since you have the oil, it will provide an appropriate 'place' for this purpose."[4] But all the means and individual supports have to be abandoned when the being surpasses the Center of the World, the only remaining legitimate mean being intellectual intuition, that is, *Buddhi* itself, which leads directly and immediately to the absolute Center.

This being, which has passed from an individual state to a universal one, reabsorbs the spiritual influences, as well as *Buddhi*; but, ultimately it is the universal spherical vortex that is reabsorbed until turning into a unique point, which is the absolute Center, where the vibratory movement becomes extinct[5] in the absolute rest, that is, in the supreme Principle,[6] and the spiritual influences that were spread in the universal manifestation (apparently only, from a metaphysical point of

---

[1] Guénon, *Aperçus sur l'initiation*, pp. 59-60.
[2] Guénon, *Articles et comptes rendus*, Éditions Traditionnelles, 2002, p. 35. "The modern people seem especially inapt to escape the materialistic order, and when they try to do so, they cannot in any case leave the domain of life" (René Guénon, *Orient et Occident*, Éditions Didier et Richard, 1930, pp. 86-88).
[3] Guénon, *La métaphysique orientale*, p. 15.
[4] *Zohar*, I, 88 a.
[5] This extinction represents *Nirvâna* or *Fanâ el-fanâi*.
[6] Cusa, *De la docte Ignorance*, p. 149.

view) are now gathered in the Center and illumine with a silent, pure and uncreated light.[1]

---

[1] This is consistent with a Masonic formula according to which the task of the masters consists in "spreading the light and gathering what is scattered" (Guénon, *Symboles fondamentaux*, p. 301).

III

THE CENTER AS A TEMPLE

René Guénon, as we have seen, implied an equivalence between the center and the tradition: the foundation of a city, Guénon said, could symbolize the institution of a doctrine or of a traditional form,[1] and Thebes is a good example, since Amphion built it using the music of his lyre, an important instrument in Orphism and Pythagorism (relating to the science of rhythm).[2] This statement[3] has more than one meaning and René Guénon's words suggest, in a veiled manner, the very process of center's "explication" (if we are permitted to use Cusa's expression) when the center manifests itself as a temple.

Amphion is the son of Zeus and has a twin brother, Zethus, and we know how important the symbolism of twins is in various traditions. However, in this case, the brothers are not representing so much the two halves of the World's Egg, but they rather indicate spiritual authority and temporal power, since Amphion is Apollo's favourite and his art of music is spiritual, while Zethus is strong like a Titan, a hunter like Nimrod and a warrior. We discover in this myth (and when we say "myth" we understand a transmission of initiatory and spiritual knowledge) how the center projected itself through a divine activity (Amphion's work[4]), that is, how the center operates as a temple, but also how a temple functions as a

---

[1] See also Benoist 42-43.
[2] Benoist 163.
[3] Guénon, *Le Roi*, p. 90. See also our *René Guénon et le Centre du Monde*, Rose-Cross Books, 2007, p. 44.
[4] Amphion, the son of Zeus, is guided by Apollo.

The Center as a temple 51

center, since Thebes' seven-gated wall was built by Zethus also, acting as a Mason.[1] Moreover, the art of Masonry, of building a temple is a "sacerdotal art" and only after the revolt of the *Kshatriyas* did this one disappear and become replaced with the "royal art," which has survived until now.[2] Amphion represents the "sacerdotal art" and Zethus the "royal art," and we find this partition in the construction of the medieval cathedrals, where the bishop or another priest in charge (like St. Bernard de Clairveaux or Suger) was the "spiritual" architect and the Master Mason was the "royal" architect.[3]

René Guénon mentioned another important symbol of the center: the *omphalos*.[4] He said that *omphalos* (in Greek "navel"[5]) was one of the most remarkable and widely spread symbol; the best-known *omphalos* was the one in the temple at Delphi, symbolizing the spiritual center for all ancient Greece. According to an ancient Greek myth, Zeus, in order to establish the center into the world, released two eagles (representing Orient and Occident) from Mount Olympus, which flew in opposite directions and met at Delphi. Zeus marked the location with the *omphalos*, the stone which his mother Rhea had

---

[1] In religious terms, the center as a temple is the descending of the divine grace, while the temple as center is the prayer. "We ought to look upon the universal world as the highest and truest temple of God, having for its most holy place that most sacred part of the essence of all existing things, namely, the heaven. ... But the other temple is made with hands; for it was desirable not to cut short the impulses of men who were eager to bring in contributions for the objects of piety, and desirous to show their gratitude by sacrifices" (Philo, *De Specialibus Legibus*, I, 66-67).

[2] See Guénon, *Symboles fondamentaux*, p. 263. René Guénon stated that the "sacerdotal art" was the art of the medieval cathedrals' builders and of the temples' builders of the Antiquity (*Autorité spirituelle et pouvoir temporal*, Véga, 1976, p. 37).

[3] "The artist's operation is dual, in the first place intellectual or 'free' and in the second place manual and 'servile' ... the nature of the ideas to be expressed in art is predetermined by a traditional doctrine, ultimately of superhuman origin" (Ananda K. Coomaraswamy, *Christian and Oriental Philosophy of Art*, Dover Publications, 1956, p. 69).

[4] *Le Roi*, pp. 74, 76, *Écrits*, pp. 90-1. See also *René Guénon et le Centre du Monde*, p. 49. See also the present work, the end of the first chapter.

[5] "The navel of the land" (*Judges* 9:37).

wrapped in swaddling clothes to take his place and fool his father Kronos. We notice that a spiritual center cannot be but in a sacred place revealed by God; in fact, the Greek Omphalos must be Zeus' equivalent, since God is the Center.[1] Delphi, Guénon specified, was considered a *universal spiritual center*, and the Omphalos symbolized the Center of the World, being considered in Antiquity a stone descended from heaven.[2] This is another example of what the syntagm "the center as a temple" means.[3]

Delphi is in direct connection with Apollo, the god of initiation, but also the god of geometry and medicine,[4] where medicine was a "sacerdotal art"[5] and geometry was the fundamental art used in building the temple[6] but also the preparatory art for a spiritual realization (as Pythagoras and then Plato affirmed, both being related to the cult of Apollo).[7] It is no coincidence that Apollo manifested himself at Delphi and we must not see a contradiction in the fact that Zeus, as stone, became the Omphalos. The Hindu, Greek and Roman gods are nothing else but "aspects," "attributes" or "faces" of the Principle,[8] and the temples built in Antiquity for different

---

[1] *Le Roi*, p. 77. The stone, having an immutable character, was a perfect symbol; the so-called Stone Age, as well as the myths regarding the people born of stones, alluded to a spiritual reality (Benoist 41).
[2] René Guénon, *Mélanges*, Gallimard, 1976, pp. 53, 55.
[3] The material representation of the *omphalos* usually consisted of a sacred stone, a *betyl*, that is, a "house of God" (Hebrew *Beith-El*), and this stone was the true "divine habitation" (*mishkan*), the seat of *Shekinah*, a designation that was later used to name the Tabernacle. Regarding the Tabernacle, this seems to be another example of a center manifested as temple, since God asked and explained to Moses how to build the Tabernacle (*Exodus* 25), He chose the architects (Bezaleel and Aholiab) and filled them with His spirit (*Exodus* 31), but, in fact, we have here an essential illustration of the temple as center.
[4] Guénon, *Mélanges*, p. 53.
[5] René Guénon, *Comptes rendus*, Éditions Traditionnelles, 1973, p. 40.
[6] Therefore Masonry was identified with Geometry.
[7] 'Abd-al-Halim Mahmûd, *Un soufi d'Occident René Guénon*, GEBO / Albouraq, 2007, pp. 125-6.
[8] As we already said, Nicholas of Cusa was the one who used the word "principle" to define the one and only God, the principle of the many gods of various people (Nicolas de Cues, *La paix de la foi*, Centre d'Études de la

deities corresponded to these various hierarchical aspects and functions necessary in supporting the evolvement of the world, each temple sheltering different spiritual influences and sometimes the same "sacred place" containing diverse temples.[1] In a traditional society, all activities, gestures, words, functions, métiers, etc. were sacred, directly related to the divine archetypes, supported by spiritual influences and supervised by gods.[2] Each temple as a house of god assisted the man in living a traditional life, but the temple facilitated also the communication with the degrees of the Universal Existence, with the super-individual states of being, allowing participation in universal harmony.[3]

Renaissance, Université de Sherbrooke, 1977, p. 45). He also used the expression "non-being," so important to René Guénon: "the All Mighty, because of his Unity, made all things to emerge from the non-being (*vocat de non esse*)" (p. 51).

[1] "We need hardly say that such a multiplicity of Gods – 'tens and thousands' – is not polytheism, for all are the angelic subjects of the Supreme Deity from whom they originate and in whom, as we are so often reminded, they again 'become one.' (…) These Gods are Angels, or as Philo calls them, the Ideas – i.e. Eternal Reasons" (Ananda K. Coomaraswamy, *What is Civilisation?*, Lindisfarne Press, 1989, p. 10); René Guénon quoted this text in *Mélanges*, p. 30. There is a subtle hierarchy of these "gods" and we should underline René Guénon's statement about the hierarchy of the spiritual centers and Ibn 'Arabî's sayings: "each of them has a piece of truth, just as each of the prophets before the Muhammadan era follows his revealed law and path"; and also: "Al-Khadir said to Moses: 'I possess knowledge that God has taught me that you do not have, and you have knowledge taught to you by God that I do not have'" (*The Meccan Revelations*, II, pp. 237, 239). However, the Prophet Muhammad appears as the seal of all the prophets: "Not so! but (we follow) the faith of Abraham the Hanîf, he was not of the idolaters. Say ye, We believe in God, and what has been revealed to us, and what has been revealed to Abraham, and Ishmael, and Isaac, and Jacob, and the Tribes, and what was brought to Moses and Jesus, and what was brought unto the Prophets from their Lord" (*Qur'an* 2:135-136).

[2] See Guénon, *Mélanges*, p. 72, where he was saying that in a traditional society, like the Islamic civilization or the medieval Christian civilization, even the most common acts had a religious character.

[3] We should not forget that a "sacred place" was also an oracle, and the Delphic oracle was eminent.

The manifestation of a center as temple implies a sacrifice as well,[1] and Guénon etymologically connected the two with the idea of cutting, separating the sacred place from the profane one,[2] in which case *templum* is "a sacred place" and *sacrum facere* (related to *sacrificium*, "sacrifice") means to make something holy or sacred. Any divine "explication" is a sacrifice, since the "complication" is incommensurably better than the "explication," but also because in this process the world is made sacred. When the Center unveils itself into the world as a temple, as a "sacred area," a sacrifice takes place, and in the case of Delphi, Apollo sacrificed the dragon Python, near the Castalian Spring. A spiritual center was usually marked not only by a sacred stone, but also by a holy spring (or fountain)[3] and by a sacred tree: at Delphi, there were the Castalian Spring (used for ritual purifications) and the laurel tree brought by Apollo from the Vale of Tempe.[4] Yet, at one moment in the evolvement of the cycle, it became necessary to point out in a more explicit manner the spiritual center, an "explication" that was in accord with the mentality of the people belonging to the last *yugas*, helping them more efficiently to follow a spiritual path and a traditional life. This is the *raison d'être* of the temple as center, and at Delphi the temple of Apollo, made of laurel, beeswax, and bronze, had been famous.

Trophonius was the mythical architect who built Apollo's temple at Delphi, who also beheaded his brother Agamedes (the necessary sacrifice) before establishing the oracle in his

---

[1] The "adverse" forces reversed the "sacrifice" and "oblation" into "corruption" and "bribery." This subject deserves a chapter by itself.

[2] Guénon, *Aperçus sur l'initiation*, p. 126.

[3] As Guénon said, in the Judaic tradition, *Malkuth*, the reservoir of the celestial waters, is identical to the spiritual center of our world (René Guénon, *Formes traditionnelles et cycles cosmiques*, Gallimard, Paris, 1980, p. 103).

[4] "The close connection between tree and sanctuary is well known; nearly every sanctuary in Palestine possessed its sacred tree and often it was the tree which determined the character of the place. ... In the Apocalypse of Moses it is said that the throne of God was fixed where the tree of life was" (A. J. Wensinck, *Tree and Birds as Cosmological Symbols in Western Asia*, Johannes Müller, Amsterdam, 1921, p. 33).

illustrious cave at Lebadeia.¹ Trophonius was considered the son of Apollo (as Apollonius of Tyana, among others, declared without any hesitation) or the son of Erginus, and he is the prototype of a Master Mason. The fact that he established an oracle makes him a messenger of the gods and explains what gave him the authority to build Apollo's temple.² However, it is said that prior to Apollo's temple there was at Delphi a more ancient sanctuary of another deity, which is a transparent allusion to the doctrine of the cosmic cycles, implying at the same time that a spiritual center develops with the passing of the *Manvantara*.

As René Guénon specified, "if we consider the history of humanity with respect to the traditional doctrines, in compliance with the cyclical laws, we must say that, originally, the man, in full possession of his state of existence, had, naturally, the possibilities corresponding to all functions, previously to any distinction of those ones," and this situation corresponded to the Earthly Paradise as One Center. "The division of these functions occurred in a later stage, representing a state already inferior in comparison to the 'primordial state,' but in which each human being, while having only some determined possibilities, had still spontaneously the effective conscience of these possibilities"; in this stage, the Center manifested itself into the world as secondary centers, and then as temples. "It is only in a period of greater obscurity that this awareness was lost; and, then, the initiation became

---

[1] See our *René Guénon et le Centre du Monde*, p. 53. "At Lebadeia is situated an oracle of Zeus Trophonius. The oracle has a descent into the earth consisting of an underground chasm; and the person who consults the oracle descends into it himself" (Strabo, *Geography* 9.2.38). Pausanias described the cave as having the shape of a bread-oven.

[2] "Obeying the oracle [of Delphi] he [King Erginus] took to himself a young wife and had children, Trophonius and Agamedes. Trophonius is said to have been a son of Apollo, not of Erginus. This I am inclined to believe, as does everyone who has gone to Trophonius to inquire of his oracle. They say that the brothers, when they grew up, proved clever at building sanctuaries for the gods and palaces for men" (Pausanias, *Description of Greece* 9. 37. 4). We may notice the two types of buildings: temples and palaces.

necessary to allow a man to find, with this consciousness, the former state to which that one was inherent"[1]: the process of building temples as centers represents this stage.

Undoubtedly, at the beginning of the *Manvantara*, the center had no image, no representation, since it did not need it. Primordial metaphysical thinking did not accept anything discursive and analytical; everything was a perfect, super-luminous and silent identity. When the Wheel started to turn and the cycle commenced to evolve, the center became ornamented with spatial and temporal coordinates, yet, even so, at the beginning, the center did not need too much of an explanation, the same way the symbols were non-explicit, synthetic and without commentaries.[2] In the primordial ages, there was no conscious idea of building a temple or a palace to symbolize the center and to be the "house of God." In those ages, it was obvious that the Center, which is the origin of the world, projected itself into the world to support it and give it reality, and the only problem was to identify such sacred places, such sacred knots, such divine projections, projections where the communication with Heaven was at hand.[3] René Guénon said: "There are places which are more capable of serving as 'supports' for the activity of the spiritual influences, and on this is based always the establishment of some principal and secondary traditional 'centers,' where the 'oracles' of Antiquity and the places of pilgrimage are the most exterior and apparent examples."[4] And René Guénon, when he explained how the

---

[1] Guénon, *Mélanges*, pp. 76-77.
[2] "At the beginning, the Book [Yi Jing] was a series of trigrams, essential and compressed ('complicated') symbols, the vision of which allowed the realization of everlasting Truth; then, the cosmic decline forced an explication and explicit description of these symbols, thus the succinct forms were born; eventually, the escalation of spiritual decadence imposed more and more elaborated commentaries" (our work *About the Yi Jing*, Rose-Cross Books, 2006, p. 8).
[3] In accord with the doctrine concerning the multiple states of the being and the multiple degrees of the Universal Existence, the sacred places could correspond to various states or degrees, and therefore, later, the pilgrimage symbolized a spiritual voyage from center to center.
[4] *Le règne de la quantité et les signes des temps*, Gallimard, 1970, p. 183.

The Center as a temple 57

Adversary tries to take possession of places which were once residences of ancient spiritual centers, added: "the reason is not only because of the psychical influences accumulated there and which are 'available'; it is also because the particular situation of these places, since it is understood that they weren't arbitrary chosen for the role that was assigned to them in one age or another and with respect to a traditional form or another."[1] "No one, O my God, changeth thy holy place; and it is not (possible) that he should change it and put it in another place: because he hath no power over it; for thy sanctuary thou hast designed before thou didst make (other) places; that which is the elder shall not be altered by those that are younger than itself."[2]

There is a "sacred geography" (about which the modern world knows nothing), and therefore, not any stone or tree or cave could represent the center as temple[3]; yet indeed, at the beginning, a tree,[4] a mountain or a cave were such temples: the paradisiacal garden and the tree of life were nothing else, and only at the end of the cycle did the center have to be presented as a built temple, in the same way Heavenly Jerusalem has been described.

René Guénon was saying that "strangely enough, in India, it is impossible to find a monument that originated before the age of Buddhism; for that reason the orientalists, who wanted to make everything start with Buddhism, tried to take advantage; yet the explanation of this fact is simple: all the previous constructions were made of wood, and obviously they disappeared without a trace."[5] Wood belongs to the vegetal symbolism of the Earthly Paradise, while the stone, as "building

---

[1] *Le règne*, pp. 252-3.
[2] *The Odes of Solomon* 4:1-4.
[3] The science of "orientation" derives from the above considerations and it was essential for the process of building a temple or a city. As Guénon said, the space is qualified by the six directions, which alludes to a profound significance of the Geometry and of the "orientation" rites (*Le règne*, pp. 53-4).
[4] Even today such trees could be seen in Varanasi.
[5] René Guénon, *La crise du monde moderne*, Gallimard, 1975, p. 23, Guénon, *Études sur la Franc-Maçonnerie*, II, p. 9, Benoist 37.

material," suggests the "solidification" of the world, even though there are other important aspects of the stone's significance.[1] From the evolvement of the cycle's perspective, we could say that the center as temple was "made," at the beginning, of the most "perishable materials" and of the least solid ones, while in modern times the builder is obsessed with the idea of a "materialistic" edifice that must stand forever: it represents a journey from invisible to visible.

However, even if it had been possible to surpass some "temporal barriers," it would be unattainable to find a temple of the earlier ages, not only because of the "perishable materials" used, but also because of what then represented a temple. For example, the holy mountain Arunâchala, which is in its essence Shiva – the Center manifested as temple,[2] began as a column, almost invisible, of light, which became a mountain of diamond in the following ages[3] and, eventually, the stony mountain we see today. The great Ramana Maharshi regarded Arunâchala not only as Shiva the Center, but also as temple, encouraging the pilgrims to ritually circle the mountain (*pradakshina*), the same way as a temple is ritually circled.[4]

In Greece, beside Delphi, Dodona has immemorially been a spiritual center and an oracle, where at one moment Zeus' oak tree (which was also an oracle), with a spring below it, became

---

[1] For Bede, wood is better than stone: the New Law corresponds to the cross made of wood [Jesus was the son of the carpenter] and the Old Law to the stone tablets of Moses; stone and wood harmoniously coexist in the temple made of stone walls and planks of cedar (pp. 41-42). In the Hindu tradition, Vishwakarma, the "Grand Architect," was called Tvashtri, the "Carpenter." As Coomaraswamy said, "'harmony' was first of all a carpenter's word meaning 'joinery,' and that it was inevitable, equally in the Greek and the Indian traditions, that the Father and the Son should have been 'carpenters'" (Coomaraswamy, *Christian and Oriental Philosophy of Art*, p. 21).

[2] "As the moon derives its light from the sun, so other holy places shall derive their sanctity from Arunâchala," Shiva declared.

[3] In various traditions, *Axis Mundi* was a diamond axis (*vajra*).

[4] The Arunachaleswar Shiva temple, built between the 10th and 16th centuries, represents the temple as center. Another famous temple dedicated to Shiva is Kashi Vishwanath temple, in holy Varanasi (considered the oldest city of the world), but nobody could discover the primary temple.

the symbol of the center as temple.¹ The cathedral of Chartres was built on a hill under which there was a cave and a spring,² which means that before the temple was founded as center in this sacred place the center functioned as temple. *The Epic of Gilgamesh* describes how the hero, at the end of his voyage through the darkness, reached "the garden of gods" where he found the divine Tree, the Tree of Life: "cornelian it bears as its fruit, bunches depend from it, beautiful to the eye; lapislazuli it bears as its twigs, fruit it bears desirable to sight."³ The tree is adorned in a similar way to which Heavenly Jerusalem is, which suggests their equivalence as temples.⁴ As we previously have seen,⁵ Poseidon built Atlantis on a mountain located in the center of an island, establishing the center in this special location: "in the center was a holy temple dedicated to Cleito and Poseidon, which remained inaccessible, and was surrounded by an enclosure of gold." The equivalence with Heavenly Jerusalem is obvious, yet this inaccessible temple is in fact the "center as temple." Later on, a temple as center was erected, very similar not only to Heavenly Jerusalem but also to Solomon's Temple: "Here was Poseidon's own temple (…) All

---

[1] "Two black doves had come flying from Thebes in Egypt, one to Libya and one to Dodona; the latter settled on an oak tree, and there uttered human speech, declaring that a place of divination from Zeus must be made there; the people of Dodona understood that the message was divine, and therefore established the oracular shrine" (Herodotus, *Histories* 2:55). The black dove is the spiritual influence or *Shekinah* and Thebes was a famous spiritual center (Guénon, *Le Roi*, p. 90). Similarly, Callimachus described in his *Hymn* dedicated to Apollo that the god, changed into a raven, showed Battos where to build the city Kyrene.

[2] Titus Burckhardt, *Chartres and the Birth of the Cathedral*, Golgonooza Press, 1995, p. 19.

[3] The ninth tablet. See *The Epic of Gilgamesh*, Penguin Books, 1968, p. 97 and Wensinck, *Tree and Birds*, 1921, p. 3. See also Philostratus: "And as to the golden olive of Pygmalion, it too is preserved in the temple of Heracles, and it excited their admiration by the clever way in which the branch work was imitated; and they were still more astonished at its fruit, for this teemed with emeralds" (*Life of Apollonius*, 5.5); the olive is a "tree of light."

[4] The "visited House," that is, the Heavenly Kaaba, is described as a building on four pillars of emerald, crowned with a hyacinth.

[5] *René Guénon et le Centre du Monde*, pp. 51-2.

the outside of the temple, with the exception of the pinnacles, they covered with silver, and the pinnacles with gold. In the interior of the temple the roof was of ivory, curiously wrought everywhere with gold and silver and orichalcum; and all the other parts, the walls and pillars and floor, they coated with orichalcum. In the temple they placed statues of gold: there was the god himself standing in a chariot – the charioteer of six winged horses – and of such a size that he touched the roof of the building with his head[1]; (...) There was an altar too, which in size and workmanship corresponded to this magnificence. (...) In the next place, they had fountains, one of cold and another of hot water."[2]

Plato also described how Poseidon, to protect Atlantis, enclosed the city with two circular ramparts and three circular fosses, alternating land and water: "Now the largest of the zones into which a passage was cut from the sea was three stades in breadth, and the zone of land which came next of equal breadth; but the next two zones, the one of water, the other of land, were two stades, and the one which surrounded the central island was a stadium only in width. (...) The entire circuit of the wall, which went round the outermost zone, they covered with a coating of brass, and the circuit of the next wall they coated with tin, and the third, which encompassed the citadel, flashed with the red light of orichalcum."

As René Guénon said, the whole world was in fact surrounded by a protective wall. The Sanskrit word *loka*, "world," comes from the radical *lok*, "to see," and is in direct relation with the light (Latin *lux*); in Latin, from *loka* was derived *locus*, "location," and Guénon considered the Masonic Lodge related to *locus*, since the Lodge is a symbol of the world and there is also the possibility of an etymological connection between *lodge* and *loka*.[3] The world inside the wall is a luminous

---

[1] In the Islamic tradition, Adam was described as gigantic as Poseidon, both representing the Universal Man.
[2] Plato, *Critias*.
[3] Guénon, *Aperçus sur l'initiation*, pp. 289-290, *La Grande Triade*, Gallimard, 1980, p. 139. The Chinese *Ming-Tang*, which was at the same time a temple

and regulated place, it is "order" (Greek *cosmos*), which became so by following God's order: *Fiat Lux*. The Great Wall encompasses this world and protects order against the malefic influences of the inferior subtle domain.¹ Indeed, the word "wall" derives from Latin *vallum*, "rampart"; thus, the primitive function of the wall was to protect and defend,² yet it is primarily not a physical protection, but a rather spiritual one, similar to the Cherubim's role to protect the Garden of Eden. Sometimes, the wall is a substitute for the city and we mentioned how Amphion, using the sounds of his divine lyre to move the stones, erected the walls of Thebes³; Poseidon built the walls of Troy, and Apollo, the god of spiritual wisdom, helped him. The wall is a "cutting," as a result of a divine activity, with two sides: one side, directed inwards, is bright and full of child-like simplicity and clarity; the other side, directed outwards, is tenebrous and intricate. On the other hand, the "cutting" makes the *templum*, that is, the "holy place" (city, fortress, palace, garden, etc.) inaccessible and inviolable for the unqualified ones, which is what became of Atlantis and the Garden of Eden.⁴

---

and a palace, was called "the Temple of Light." The Lodge is also "the most luminous place." In Romanian, the words *lume* ("world") and *lumina* ("light") are very closely related.

¹ Guénon, *Le règne*, p. 230. In the Hindu tradition, the Great Wall is the circular mountain *Lokâloka*, which separates the "cosmos" (*loka*) from the outside darkness (*aloka*).

² The root *var* or *vri*, which has produced the Sanskrit words *varâha* and *vrika*, means also "to cover, to protect, enclosure, walls" (Guénon, *Symboles fondamentaux*, p. 179). In Iranian tradition, Ahura Mazda asks Yima to build a *Vara*, i.e., an "enclosure" to preserve the germs of all the animals and plants, *Vara* being similar to Noah's Ark. Sanskrit and Iranian *vara* is close to Latin *vallum* (to be compared with Sanskrit *sûrya* and Latin *sol* designating the sun). The Celtic *var* means "fortified city"; from this word the Hungarian *varos* and Romanian *oras* derive, with the same meaning.

³ See also W. F. Jackson Knight, *Vergil, Epic and Anthropology*, Barnes & Nobles Inc, 1967, pp 118-119 (see, for example, René Guénon's opinion about Jackson Knight in his article *La Caverne et le Labyrinthe*, *Symboles fondamentaux*, p. 209).

⁴ See Knight. Knight concluded that: the cities in early Greece and Italy were holy and defended by holy walls; they were sometimes closely identified with

René Guénon illustrated how a desolate center changes from the "city of God" to a "city of Devil" using an Islamic concept about *les fissures de la Grande Muraille*, "the fissures in the Great Wall."[1] This Wall is in the Islamic tradition the "rampart" that protects against Gog and Magog[2] and many *hadîth* talk about the crack in the wall, which will bring the end of the world.[3]

Of course, there were people who asked again and again how could God build a penetrable Wall, which is equivalent to asking how evil can exist? Agarttha, as center, was impenetrable and inviolable; Poseidon's temple was inaccessible. Then how could these centers be invaded by the counter-initiatory forces?[4] In fact, this question is wrongly put, and the answer is well known for the pupils of Tradition.

Nicholas of Cusa declared with much fervour: "While I imagine a Creator creating, I am still on this side of the wall of Paradise! While I imagine a Creator as creatable, I have not yet entered, but I am in the wall. But when I behold Thee as Absolute Infinity, to whom is befitting neither the name of

---

an important temple; they were sometimes closely identified with goddesses of the cities' defence; the goddesses may wear a *polos*, or towered crown, indicating that these personalities are themselves peculiarly defended by the city walls (see also Philip Sherrard, *Byzantium*, Time-Life Books, 1975, p. 15, where 5th Century ivory plaques are presented, showing the old capital, Rome, as a woman wearing a military helmet, and the new capital, Constantinople, as a woman wearing a crown symbolizing the walls of the city); the walls of cities and buildings are sacred (pp. 293-294). In fact, Knight's conclusions could be summarized by saying that the center is not different from god, and its images are cities, temples or *palladiums* (at one moment, Constantinople's *palladium* was Hodegetria, the icon of the Virgin Mary, painted by St. Luke); also, the king wearing a crown should represent the very "holy land" defended by the sacred walls.

[1] Guénon, *Le règne*, p. 230 ff.
[2] *Qur'an* 18:92-99.
[3] Émile Dermenghem, *Muhammad*, Harper & Bros. 1958, pp. 20-1.
[4] See Knight 110-111. Knight stressed the sacredness of the walls, which in the case of Troy (a secondary center) were called "the sacral veil"; Homer, describing the building of the Trojan walls by the gods, said that their intention was "that the *polis* might be unbreakable," yet the crack in the wall was made by the Trojan horse (Quintus Smyrnaeus said that the Trojans introduced the horse by "releasing the sacral veil of their great city").

creating Creator nor of creatable Creator – then indeed I begin to behold Thee unveiled, and to enter in to the garden of delights!"[1] For Cusa, the Center (which is Paradise) is surrounded by a wall that represents *coincidentia oppositorum*. Using Guénon's metaphysical language we could say that outside the wall there are the possibilities of manifestation that manifest themselves (the Creator creating), within the wall there are the possibilities of manifestation that don't manifest themselves (the creatable Creator), and inside the wall there are the possibilities of non-manifestation; or, even closer to what Cusa was describing, we could say that the creating Creator is the Being in manifestation, the creatable Creator is the Being in itself, non-manifested, and the Non-Being is inside the wall, in the Center.

We see the difference in comparison to the Great Wall that supposes to envelope the whole world. In the case of Atlantis, only a traditional form was protected by the sacred wall; in Cusa's description, only the very Center is guarded by the wall. However, as we explained in our first chapter, in a normal society the whole world should be a holy land emanating from the center, with the specification that a symbolic hierarchy of the spiritual Pole's successive approximations should be observed.[2] Yet, even if we accept the world as a complete "holy land," a *templum*, only the Center is immutable and imperishable[3]; the possibilities of manifestation that manifest themselves must follow the laws of manifestation, of becoming, that is, they obey to the law of the cosmic cycles. "The Principle," Zhuang Zi stated, "is immutable, without beginning or end. The beings are changeable; they are born and die,

---

[1] Nicholas of Cusa, *The Vision of God*, The Book Tree, 1999, p. 57.
[2] The Zodiac symbolizes the wall of the Cosmos. In the well known image of Shiva Natarâja, the cosmic wall is symbolized by a circle of fire and light (*tiruvâsi* or *prabhâ-mandalâ*).
[3] In various traditions it is stated how the center was manifested first and then the world. In the *Odes of Solomon* (4:1-4): "thy sanctuary thou hast designed before thou didst make other places"; in the Islamic tradition, Kaaba was created before the rest of the world (A. J. Wensinck, *The Navel of the Earth*, Johannes Müller, Amsterdam, 1916, p. 17-19).

without permanence." In the *Yi Jing*, "change" has more than one meaning: the main sense, *bu-yi*, "the unchangeable change," corresponds to the supreme Void; the cosmologic sense, *jian-yi*, "the change," corresponds to the Being; the worldly sense, *bian-yi*, "the changeable change," corresponds to the multiplicity and the contingencies of the world. As Saint John Damascene said: "All the existent things are created or uncreated; if they are created, they are without doubt also changeable, either by destroying themselves, or by changing in a free mode. If they are uncreated, they are without doubt also unchangeable; ... even the angels change, suffer modifications. (...) But the Creator has to be uncreated. Therefore, the Creator being uncreated, without doubt He is also unchangeable." And Muhhyiddin Ibn 'Arabî stated: "Allâh is the only one that has the power to produce the essences and to modify the states. (...) The beings are doomed to disappear, to appear, to change, to be destroyed and to pass from a situation to another. They exist due to 'one other than themselves' and the existence is for them a loan, a transfer. (...) Without doubt, everything is illusory but Allâh. Everything is changeable and perishable but His Face."[1]

Thus, the world is doomed to decay, to suffer the changeable change, and, at one moment, the malefic forces will find no Masons ready to repair the fissures, and the adversary will penetrate the wall. If, at the beginning, due to the Great Wall the world as a *templum* was impenetrable and inaccessible for the agents of outside darkness, at the same time it was completely open, penetrable, and accessible for the spiritual influences from above. Along with the solidification of the world, the "shell" built by materialism started to block any communication with the superior Center,[2] which has created an upside-down situation with a world that became more and more inaccessible for spiritual influences and increasingly defenceless in the face of inferior forces: we recognize here the parody that dominates the end of time. As René Guénon said, the action of

---

[1] See our work *About the Yi Jing*, Rose-Cross Books, 2006, pp. 40-41.
[2] Guénon, *Le règne*, p. 231.

the spiritual centers became more and more closed, since the spiritual influences that were normally transmitted to our world could not manifest themselves outward, being stopped by the impenetrable "shell."[1]

The last statement gives us the opportunity to further clarify the symbolism of the center and to stress the difference between the supreme Center and the secondary centers. A secondary center is a "house of God," but in fact only the supreme Center is the *real* House of God and only there God, the Center, and the Tradition (or the Primordial Tradition) are one and the same.[2] As René Guénon said, the secondary center (that is, the center of a traditional form) is an emanation or a reflection (an image) of the supreme spiritual Center, with which it is virtually identical, and therefore the land containing such a center is a "holy land," while the traditional form attached to it communicates through this very secondary center with the supreme Center.[3] The supreme Center transmits the spiritual influences to the secondary center, which means, using the language of the Judaic Kabbalah, that the place where the secondary center is established is the location of *Shekinah*'s manifestation, or, in other words, that the secondary center becomes the house of the "divine presence," and René Guénon gave as example the Temple of Jerusalem, which, for the Hebrew tradition, was such a secondary center. Even when the tradition (the regular and orthodox transmission) is provided by

---

[1] Guénon, *Le règne*, p. 234. This upside-down situation highlights the "iron wall" that isolated the world from heavenly activity: "Since the day that the Temple was destroyed, an iron wall has intervened between Israel and their father in heaven" (*Berakhot* 32b); "But your iniquities have separated between you and your God, and your sins have hid his face from you, that he will not hear" (*Isaiah* 59:2).

[2] However, a secondary center, as an image of the Center, is virtually identical to God. Therefore, Mohyiddin Ibn 'Arabî could affirm that the inspiration to write *Al-Futûhât al-Makkîya* ("The Meccan Revelations") came to him during his first pilgrimage: "The essence of what is included in this work comes from what God inspired in me while I was fulfilling my circumambulations of His Temple [the Kaaba], or while I was contemplating it while seated in its holy precincts" (Ibn Al 'Arabî, *The Meccan Revelations*, Pir Press, 2005, I, p. 8).

[3] Guénon, *La Grande Triade*, p. 138, *Aperçus sur l'initiation*, p. 65.

a secondary center, this must be in relation to the supreme Center – the immutable keeper of the Primordial Tradition that represents the origin of all the particular traditional forms, and it has to be stressed that the attachment of the secondary center to the supreme Center is indispensable to maintain continuity in the transmission of the spiritual influences from the origin of present humanity, where this origin is, in fact, the supreme Center.[1]

There is a hierarchy of the spiritual centers,[2] which allows René Guénon to say that Seth, the elected one who was able to return to the Terrestrial Paradise (the Center) and recover the Holy Grail (a deed representing the restoration of the primordial order), could establish a spiritual center replacing the lost Paradise and being its image, since the possession of the Holy Grail means the integral conservation of the Primordial Tradition in such a spiritual center[3]; of course, this type of center is qualitatively and hierarchically much more than a secondary center related to a particular traditional form. Losing the Holy Grail symbolizes, like the Lost Word, the loss of the Primordial Tradition, but, as we already said, this can happen only to a particular tradition and in the case of the secondary centers, while the supreme Center or its substitute will keep untouched the repository of the Tradition, which will become rather hidden than lost.[4]

Therefore, we can say that in the *Kali-yuga* the supreme Center became hidden, subterranean, and closed, while the number of the secondary centers representing it externally

---

[1] See Guénon, *Aperçus sur l'initiation*, p. 65.
[2] "An initiatory organization could derive from the supreme Center, not directly, but through the intermediary secondary and subordinated centers. ... As there is a hierarchy of degrees in each organization, so there is among the organizations themselves" (Guénon, *Aperçus sur l'initiation*, p. 67).
[3] There is an Islamic tradition about the "treasure of Adam": God sent down to Adam a chest (ark, *tâbût*) containing the images of all the prophets from Adam to Muhammad; after Adam's death the chest was transmitted to Seth and eventually it became Moses' ark (Michel Vâlsan, Études Traditionnelles, 1962, no. 374). We note the correspondence with the Holy Grail.
[4] Guénon, *Le Roi*, p. 43.

decreased dramatically.¹ We envision two situations: first, when a traditional form degenerated and became extinct, the spiritual influences abandoned the secondary center and withdrew into the supreme Center, leaving behind psychical elements that could be used by the counter-initiatory forces which penetrated the Great Wall and occupied the deserted center²; second, the wall of the secondary center or of the "holy land" has been broken and the adversary infested the corresponding traditional organization.³

The supreme Center itself is described as surrounded by a wall. Nicholas of Cusa described the Paradise: "the place wherein Thou art found unveiled is girt round with *coincidentia oppositorum*, and this is the wall of Paradise wherein Thou dost abide."⁴ "For Thou art separated by an exceeding high wall from all these. The high wall separates Thee from all that can possibly be said or thought of Thee, forasmuch as Thou art Absolute above all the concepts which any man can frame."⁵

The importance of the wall's symbolism explains why the center as temple was sometimes described as being a fortress or a city, and we stress again that *templum* means primarily a "holy place."⁶ It is well known how St. Augustine has described two

---

¹ Guénon, *Le Roi*, pp. 70-71. Some ignorant detractors tried to demolish René Guénon's sayings about Agarttha, the subterranean center, suggesting that such an idea cannot be found in the Hindu tradition. There is in Haridwâr a *Shiva Lingam*, which naturally emerged, and which, with the evolution of the cycle, progressively retracted underground. Today, you can see just its top, since it is the end of the *Kali-yuga* and the center is almost completely subterranean.
² Guénon, *Le règne*, p. 248.
³ Guénon, *Le règne*, p. 252.
⁴ Cusa, *The Vision of God*, pp. 44, 46-47, 49, 50, 53.
⁵ Cusa, *The Vision of God*, p. 59. In the Eastern Orthodox Church, this wall is the *iconostasis* that protects the Holy of Holies or the Altar.
⁶ The center of the city was usually marked by the temple and/or the royal palace as "the center of the center" or "the heart of the city." Guénon specified that, originally, the palace was a temple too (*Symboles fondamentaux*, p. 451). Any spiritual center could be described as a temple (the sacerdotal aspect) and as a palace (the royal aspect) (Guénon, *Le Roi*, p. 26), or both (like the *Ming-Tang*) (Guénon, *Symboles fondamentaux*, p. 451).

cities, the city of God and the terrestrial city (which is not a spiritual center),[1] based on the psalmist's words: "Glorious things are spoken of thee, O city of God"[2]; "Great is the Lord, and greatly to be praised in the city of our God, in the mountain of his holiness."[3] Even though the "city of God" mentioned in the *Psalms* is usually considered to be Jerusalem, St. Augustine specified: "we have learnt that there is a City of God: and we have longed to become citizens of that City... But the citizens of the earthly city prefer their own gods to the founder of the Holy City, not knowing that he is the God of gods."[4] The founder of the Holy City is also the builder or the Great Architect: "Every house is built by someone, of course; but God built everything that exists,"[5] and for this very reason the symbolism of the temple as center is perfectly legitimate and orthodox; and He is also the only Citizen.

In the Islamic tradition, each spiritual station is a city as "holy place," yet what makes the city holy is not the place itself but the Citizen or its representatives, the holy citizens. Ibn 'Arabî explained: "This (power of spiritual places) is due to those who inhabit that place, either in the present, such as some of the noble angels or the pious spirits (*djinn*), or else through the spiritual intentions (*himma*) of those who used to inhabit them ... those influences have remained behind them, so that sensitive hearts are influenced by them. ... The knowledge of the spiritual influence of places and the sensitivity to its greater or lesser presence is part of the completion of the mastery of the Knower."[6]

From an absolute metaphysical perspective there are no "citizens" of the divine City, but only one "citizen." Ananda K. Coomaraswamy, who developed this theme,[7] quoted Philo: "As

---

[1] St. Augustine, *City of God*, Penguin Books, 1984, p. 593.
[2] *Ps.* 87:3.
[3] *Ps.* 48:1.
[4] *City of God*, p. 429.
[5] *Hebr.* 3:4.
[6] James Winston Morris, *The Reflective Heart*, Fons Vitae, 2005, pp. 66-67.
[7] See Ananda K. Coomaraswamy, *What is Civilisation?*, Lindisfarne Press, 1989, 1989, Guénon, *Comptes rendus*, p. 181.

The Center as a temple 69

"As for lordship, God is the only citizen." In the Hindu tradition: "This Man (*purusha*) is the citizen (*purushaya*) in every city."[1] René Guénon, at the end of his life, wrote an article called *La Cité divine* ("the Divine City"), where he used Ananda K. Coomaraswamy's study, showing how the symbolism of the "City of God" is open to more than one "macrocosmic" application at different levels: the City could be the center of a world (of a particular state of existence), corresponding to Heavenly Jerusalem, or it could be the center of all the worlds (which represent the whole of the universal manifestation).[2] In both cases, there is in this center a divine Principle (*Purusha*,[3] equivalent to *Spiritus Mundi* of the Western traditions), defined in the *Satapatha Brâhmana*: "And as to why it is called Purushamedha: The fortress or city (*pur*) is without any doubt these worlds, and Purusha is he that blows here, he bides in this city (*pur*): hence he is Purusha. And whatever food there is in these worlds that is its *medha*, its food; and inasmuch as this is its *medha*, therefore it is called Purushamedha. And inasmuch as at this sacrifice he seizes men (*purusha*) meet for sacrifice (*medhya*), therefore also it is called Purushamedha" (XIII.6.2.1).

As the Hindu city has *Purusha* its only citizen, as the *stûpa* is Buddha's body, so the Christian church is Christ's body, His archetypal sacrifice allowing, in a way, the raising of the Church.[4] Eusebius wrote inspiring words about it: "Thus this one[5] also bearing in his own soul the image of the whole Christ, Christ, the Word, the Wisdom, the Light, has formed this

---

[1] *Brhadâranyaka Up.* II.5.18, Coomaraswamy, *What is Civilisation?*, p. 2.
[2] Guénon, *Symboles fondamentaux*, p. 452.
[3] The Hindu tradition considers *Purusha* dwelling in the Sun, which is a symbol symbol for the Center.
[4] As the Church Fathers stressed, the temple is a reflection of Christ's body and not the reverse (Burckhardt, *Chartres*, p. 21). However, Guénon preferred to say that "the humanity of Christ was compared to the temple," since the human composite and not only the body is destroyed be death (*Écrits*, p. 110); and he added that the Heart is, in his humanity, what the Holy of Holies is in the Temple.
[5] Paulinus, bishop of Tyre. See also Jean Hani, *Le Symbolisme du Temple Chrétien, Chrétien*, Guy Trédaniel, 1978, p. 30.

magnificent temple of the highest God, corresponding to the pattern of the greater as a visible to an invisible..."[1]; "But in the leader of all it is reasonable to suppose that Christ himself dwells in his fullness."[2]

In connection with the quote from the *Satapatha Brâhmana* alluding to the sacrifice, we should add René Guénon's statement about Christ being exclusively the victim and the priest at the same time, just as Purusha, who is identical with Prajâpati ("the Lord of the produced beings") and with Vishwakarma ("the Great Architect of the Universe"), is the one who accomplishes (as Vishwakarma) the sacrifice and is its victim too.[3] In the Christian tradition, Holy Communion is a sublime mystery of the Church through which is repeated Christ's sacrifice made at the Last Supper; yet, it is not about a sectarian act through which Jesus is "cut into pieces," divided, and his parts scattered all over the corporeal world, but a sacrifice that unites, imparts, and brings together a sacred community.[4] It illustrates the enigmatic process of multiplicity being the "explication" of One, meaning that even though, to produce the universal manifestation, *Purusha* had to be sacrificed and divided into multiple parts, in reality He was, is and will always be the same One and only, the only Citizen who dwells in all the multiple cities.[5]

In the Hindu tradition, Vastu Purusha was sacrificed by the *dêvas*, but, to suggest that its unity was not in fact damaged, it is said that the *dêvas* took possession of Purusha's various parts or

---

[1] Eusebius Pamphilus: *Church History, Life of Constantine, Oration in Praise of Constantine*, T&T Clark, Edinburgh, 1890, X.4.26.
[2] Eusebius X.4.67. Bede underlined too how the temple is the body of Christ (Bede, *On the Temple*, Liverpool Univ. Press, 1995, pp. XXVIII, 5). "In principle, the body of Christ is the model for all constructed sanctuaries" (Burckhardt, *Chartres*, p. 22). See also Titus Burckhardt, *Principes et méthodes de l'art sacré*, Dervy, 1976, p. 68 ("The symbolism of the Christian temple is based on the analogy between the temple and the body of Christ").
[3] Guénon, *Symboles fondamentaux*, pp. 301-302.
[4] See our work *The Wrath of Gods*, Rose-Cross Books, 2004, p. 18.
[5] Since the church is a reflection of Christ's body, Holy Communion is as well well a sacrifice that allows the building of the temple.

filled its body.¹ The sacrifice of Vastu Purusha (*vastu* means "existence"²) illustrates the process of the Center manifesting itself as temple, the Hindu temple being the image of Vastu Purusha, while the *Vastu-Purusha-mandala* represents the sacred diagram of the *templum*; and, we should stress that a *mandala* defines and delimits a "holy land," protecting it against the malefic forces, a "holy land" that from a universal perspective is the whole Cosmos.³ The center of this "holy land" is, in the case of the *Vastu-Purusha-mandala*, the central square called *Brahmâsthana*, "the station of Brahmâ," corresponding to the Chinese *Zhong Yong*, the "Invariable Middle."⁴ Burckhardt considered Ayodhyâ, Râma's inaccessible and fabulous solar city,⁵ a *mandala* having in the center the "divine city" *Brahmapura*, an equivalent of *Brahmâsthana*.⁶ On the other hand, *Brahmapura* itself is characterized as *ayodhyâ*, that is, "free of conflicts," a place of everlasting harmony and peace, and we remember Cusa's description of Paradise's walls as *coincidentia oppositorum*.

---

[1] As the light allows us to see distinctly the different elements of the world, so the *dêvas* illuminates the multiplicity from One. From a *prakritian* perspective, this is *Fiat Lux*, when Vastu Purusha appears as Mûla-Prakriti, in which case *Prakriti* is the source of multiplicity, while *Purusha* remains undivided.

[2] From a metaphysical viewpoint, the sacrifice signifies the production of manifestation (*vastu*) from non-manifestation (*vustu*).

[3] The concept of the Center manifested as temple permitted the development of the sacred architecture having as objective the building of the temple as center. The *Vastu Shastra*, the scripture about the sacred architecture, is considered the oldest known architectural treatise, but before it, there was an oral architectural tradition, which goes beyond the so called historical times. About the *Vastu-Purusha-mandala* see also Burckhardt, *Principes*, pp. 33 ff.

[4] We should add that the *Vastu-Purusha-mandala* is at the same time the "architectural diagram" of the Cosmos, temple, city, royal palace and throne (Burckhardt, *Principes*, p. 44). In India, the throne was associated with the tree, both symbolizing the center as temple; the throne (*vajrasana*) is a place where the "divine presence" manifests itself.

[5] In the *Râmâyana*, Manu, the Lord of the World, is the builder of Ayodhyâ.

[6] Burckhardt, *Principes*, p. 44. At the beginning of the *Râmâyana*, Ayodhyâ is compared to a game board (*astapadi*, "eight-feet"), which is a chessboard and equivalent to the *mandala* with 64 squares.

Sometimes, *Brahmapura* is explicitly identified with the Center of the World, the Hindu tradition locating it on top of Mount *Mêru*, but usually *Brahmapura* symbolizes the center of the being, the "heart,"[1] this microcosmic perspective, in perfect accord with the macrocosmic one, describing, as in the case of the Kingdom of God,[2] how the Center operates as temple. Symbolically, as René Guénon explained, based on the Hindu tradition, the Ether is, from a physical viewpoint, the resident of the vital center; *jîvâtmâ*, "the living soul," is, from a psychical perspective, the resident of this center; and *Âtmâ*, the Self, identical with *Brahma*, is, from a metaphysical point of view, the occupant, the unique citizen, which motivates the name *Brahmapura* for the heart.[3] *Brahma*, called *Purusha* (mainly from a microcosmic view),[4] is sacrificed and divided (apparently) in order to populate all the worlds and all the beings, and as *Purusha* fills up the whole Cosmos, Temple or City (since in reality he is the only citizen), so he fills up the integral being.[5] However, recalling the spiritual Pole's successive approximations, *Brahmapura* is primarily the Center or the Heart of the universal manifestation or of the integral being,[6] and we should be able to see the hierarchic and qualitative difference between the body of Christ, having the temple (which symbolizes the center) as a reflection, and the Sacred Heart of Christ.[7] Meister Eckhart, for example, compared the temple with the soul:

---

[1] Of course, it is not only about the corporeal organ, since beyond the physical or senses' domain, we cannot talk about "location" and the "heart" has to be understood in a symbolic manner. See René Guénon, *L'homme et son devenir selon le Vêdânta*, Éd. Traditionnelles, 1991, p. 42, and *Symboles fondamentaux*, pp. 441, 443.

[2] "Behold, the Kingdom of God is within you," *Luke* 17:21. St. Symeon the New Theologian asked: "How can a person who does not possess in himself the Kingdom of Heaven in a conscious way enter into it after death?"

[3] *L'homme et son devenir*, p. 46, *Symboles fondamentaux*, p. 446.

[4] *Katha Upanishad* II.2.8 (*Eight Upanisads*, Advaita Ashrama, Calcutta, 1972, I).

[5] Guénon, *Symboles fondamentaux*, pp. 450-1.

[6] Guénon, *Symboles fondamentaux*, p. 451.

[7] If in our lower world the body contains the heart, in the higher world the heart contains the body or the center contains the integral being. Similarly, the Sacred Heart contains the temple.

The Center as a temple   73

"Jesus alone wants to speak in the temple, that is, in the soul."[1] He explained in a sermon the spiritual meaning of the "cleansing of the Temple," when Christ drove out the sellers[2]: our heart has to be clean and empty of any other passions and desires[3]; only then will God come to live in his temple, and Meister Eckhart repeatedly affirmed that the only "citizen" of the soul has to be God or Jesus.[4] Yet in the soul, which is equivalent to the temple, is the "location" of the real center, called by Meister Eckhart *bürgelin*, "fortress," or "small fortress of the soul,"[5] identical with *Brahmapura*, the "divine city," and, indeed, Eckhart said also: "this 'city,' in a spiritual sense, is the soul.[6] 'City' means *civium unitas*, that is, a closed city to the exterior and united internally."[7]

This definition of the "city" should be related to Meister Eckhart's sayings about the purified soul: "The soul is purified in the body in order to reassemble what was scattered and taken to the exterior.[8] When what the five senses take to the exterior comes back into the soul, this one has a power in which everything becomes one,"[9] which means that we should give up up God for God,[10] sacrificing the sacrificed *Purusha* for the undivided, one without a second, imperishable *Purusha*. Indeed, as René Guénon said, following the Hindu tradition, *Purusha* is

---

[1] Maître Eckhart, *Sermons*, Éd. du Seuil, 1974, I, p. 48.
[2] *John* 2:15.
[3] For Meister Eckhart, the corporeal heart is the noblest organ and located in the center of the body, *Sermons*, III, pp. 140-1.
[4] Eckhart, *Sermons*, I, pp. 46-8.
[5] *Intravit Iesus in quoddam castellum*, *Luke* 10:38, Eckhart, *Sermons*, I, pp. 50 ff. See See our work *The Everlasting Sacred Kernel*, Rose-Cross Books, 2001, p. 5.
[6] The city is the image of the soul, said St. Catherine of Siena (Titus Burckhardt, *Siena, The City of the Virgin*, Oxford Univ. Press, 1960, p. 53).
[7] Eckhart, *Sermons*, III, p. 137.
[8] We recognize here a renowned Masonic formula. See Guénon, *Symboles fondamentaux*, p. 300, where he identified symbolically "what was scattered" with the parts of sacrificed and divided *Purusha*; this "scattering" represents the passage from unity to multiplicity.
[9] Eckhart, *Sermons*, I, p. 95. We notice the accord with the Hindu tradition (see, for example, *pratyahara*).
[10] Eckhart, *Sermons*, I, p. 122.

one and multiple: "There are these two Purushas in the world, the perishable and the imperishable: the first is divided among the beings, the second is immutable."[1]

The Hindu tradition, rightly concerned with the people's capacity to err, especially in the *Kali-yuga*, stressed the symbolism of one and multiple in many ways. The image of Shiva Natarâja is a famous application of it, yet there is a perfection and symbolic fullness of this representation, which makes its significance to go far beyond this particular aspect. The dance of Shiva is performed in Chidambaram or Tillai, which is the Center of the Universe,[2] but there is also a "deeper "deeper significance," Shiva dancing in fact within our heart, in our self.[3] He has one foot planted on the ground, sustaining multiplicity, and the other one lifted in the air, which symbolizes Liberation, immutability, eternal bliss.[4] Multiplicity

---

[1] *The Bhagavad-Gîtâ*, XV.16, Guénon, *Symboles fondamentaux*, p. 451. Coomaraswamy (*Metaphysics*, p. 222) followed Dionysius the Areopagite to illustrate the coexistence of unity and multiplicity; Dionysius said: "Every number preexists uniquely in the monad and the monad holds every number in itself singularly. Every number is united in the monad; it is differentiated and pluralized only insofar as it goes forth from this one. All the radii of a circle are brought together in the unity of the center which contains all the straight lines brought together within itself. These are linked one to another because of this single point of origin and they are completely unified at this center. As they move a little away from it they are differentiated a little, and as they fall farther they are farther differentiated. That is, the closer they are to the center point, the more they are at one with it and at one with each other, and the more they travel away from it the more they are separated from each other (*The Divine Names*, 821A, *Pseudo-Dionysius, The Complete Works*, Paulist Press, 1987, pp. 99-100). For Nicholas of Cusa, "in the wall of Paradise, where Thou, my God, dwellest, plurality is one with singularity" (*The Vision of God*, p. 83).

[2] Ananda K. Coomaraswamy, *The Dance of Siva*, Dover, 1985, p. 57, Ananda K. K. Coomaraswamy and Sister Nivedita, *Myths of the Hindus and Buddhists*, Dover, 1967, p. 312.

[3] Coomaraswamy, *The Dance of Siva*, pp. 60-1, *Myths*, p. 313.

[4] One of Shiva's hands points downward to the uplifted foot and is held in a pose reminiscent of Ganesha's trunk (Heinrich Zimmer, *Myths and Symbols in Indian Art and Civilization*, Harper & Row, 1962, p. 153). The significance of Jason, the conqueror of the Golden Fleece, having one bare foot and the

is suggested by the *prabhâ-mandala*,[1] the ring of fire and light, but in one hand Shiva bears also a tongue of flame, which symbolizes the destruction of the world by fire at the end of times, in contrast to the fire that "energizes" the beings and the worlds (*Fiat Lux*); however, this encircling fire (or better, light) (*tiruvâsi*) is more than *Fiat Lux*, symbolizing transcendental light and even the sacred syllable AUM.[2] The Christian tradition stresses too how the fire has two powers: one destructive, the other illuminating; St. John of the Ladder said the same fire is called both "that which consumes and that which illuminates," since God is both illumination and hell for men.[3]

The illumination is the essence of *Brahmapura* implicitly. In the divine city, "the sun does not shine, neither do the moon and the stars; nor do these flashes of lightning shine. How can this fire? Purusha [*Brahma*] is shining all these shine; through his lustre all these are variously luminated."[4] Heavenly Jerusalem is described in a similar way: "And I saw no temple therein: for the Lord God Almighty and the Lamb are the temple of it. And the city had no need of the sun, neither of the moon, to shine in it: for the glory of God [*Shekinah*] did lighten it, and the Lamb is the light thereof."[5] "And there shall be no more curse: but the throne of God and of the Lamb shall be in it; and his servants shall serve him: And they shall see his face; and his name shall be in their foreheads. And there shall be no night there; and they need no candle, neither light of the sun; for the Lord God giveth them light: and they shall reign for ever and ever."[6]

---

other wearing a sandal should be compared to that of Shiva's feet; there is also a relation with the Masonic initiatory ritual.

[1] *Prabhâ* means "light."

[2] Zimmer, *Myths*, p. 155. Meister Eckhart described the "small sparkle of the soul (*scintilla animæ*) created by God and which is a light imprinted from above (*Sermons*, I, p. 175).

[3] Hierotheos, *Life after Death*, Birth of the Theotokos Monastery, 1998, pp. 257-261.

[4] *Khata Up.* II.2.15.

[5] *Revelation* 21:22-23.

[6] *Revelation* 22:3-5. See Guénon, *L'homme et son devenir*, pp. 47-8, *Symboles fondamentaux*, pp. 434-6, 440, 450. Guénon underlined the similitude between *Brahmapura* and the Heavenly Jerusalem. See also Eckhart: "This city did not

It is remarkable how the data from different traditional forms are in perfect harmony when describing the Center.

In the Chinese tradition, the Center is marked by the *Ming-Tang*, "the Temple of Light," composed of nine rooms, symbolizing the nine provinces, and having 12 windows (two windows for the corner rooms, and one for each middle room of the four sides),[1] which makes it comparable to Heavenly Jerusalem, as a temporal and spatial image of the Universe. As Guénon noted the resemblance between *Brahmapura* and Heavenly Jerusalem, so Titus Burckhardt observed the similarity between Heavenly Jerusalem and *Vaikuntha*, the Heavenly Center of Vishnu.[2]

Here is the description of *Vaikuntha* from the *Skanda Purana*: "Behold the Temple of Gems standing on the White Island surrounded by the Ocean of Milk. In the midst of the Milk-Ocean the Sacrificial Hall is made of precious stones. It is built of pure crystal and is unshakable. The interior of the Temple is in twelve by twelve parts and is shining with the fiery brilliance of the Sun. It is resting on sixteen pillars made of emeralds and has twelve portals towards the four directions of space. The walls of the secondary shrines in the four corners are made of rubies and have perforated windows with sixteen openings. These are sixteen parts *(kalâs)*, by adding which the full number of 64 *kalâs* is obtained.[3] The beautiful sacrificial Hall is emitting a light equal to a myriad of Suns, and that light will endure to the end of all the *kalpas*. In the center of the Hall

---

have a house of prayer, God himself was the temple. It had no need of the light of the sun, neither of the moon: the clarity of God did luminate it. This city designates every spiritual soul as St. Paul said: the soul is a temple of God." (Eckhart, *Sermons*, I, pp. 46-47, II, p. 178).

[1] Guénon, *La Grande Triade*, p. 141, Marcel Granet, *La pensée chinoise*, Albin Michel, 1968, pp. 150 ff.

[2] Titus Burckhardt, *Symboles*, Archè, 1980, pp. 29 ff. It is worth also to recall the likeness between the depiction of Atlantis and that of Heavenly Jerusalem.

[3] We remember the *mandala* of 64 squares. The traditional Hindu city was composed of 4 x 4 "quarters," separated by six paths (three oriented north-south, three east-west); the royal way or the main way connected the western gate with the eastern gate and led to the royal palace built in the center of the city.

The Center as a temple

there is the stainless Tree (of Life) arising from the shining, hundred-petaled lotus. Its roof has two stories and is covered with golden tiles. Between the stories there is a perforated wall made of pearls. On the top of the roof is a beautiful *kalasha*, a golden pitcher filled with the Milk of Immortality.[1] Two divine birds are sitting by the side of the *kalasha* in perfect silence.[2] In this self-luminous, brilliant sanctuary no sun is shining, no moon and no stars. This is the abode of *Nârâyana*, who is beyond the changeable world [the universal manifestation[3]] and beyond the unchangeable [the Supreme Being[4]]. I worship this *Purushottama*,[5] who in all the Three Worlds (*Tribhuvana*) is the most difficult to approach."

Heavenly Jerusalem is described in a most akin way: "And he carried me away in the spirit to a great and high mountain, and shewed me that great city, the holy Jerusalem, descending out of heaven from God, having the glory of God: and her light was like unto a stone most precious, even like a jasper stone, clear as crystal; and had a wall great and high, and had twelve gates, and at the gates twelve angels, and names written thereon, which are the names of the twelve tribes of the children of Israel: On the east three gates; on the north three gates; on the south three gates; and on the west three gates. And the wall of the city had twelve foundations, and in them the names of the twelve apostles of the Lamb. (...) And the building of the wall of it was of jasper: and the city was pure gold, like unto clear glass. And the foundations of the wall of the city were garnished with all manner of precious stones. The first foundation was jasper; the second, sapphire; the third, a chalcedony; the fourth, an emerald; the fifth, sardonyx; the sixth, sardius; the seventh,

---

[1] The pitcher is the equivalent of the Holy Grail, marking the center. In *Titurel*, the sanctuary of the Holy Grail is an image of Heavenly Jerusalem.
[2] The two birds recall *Âtmâ* and *jîvâtmâ* from the *Upanishads*.
[3] *Bian-yi*, "the changeable change," the perishable *Purusha*.
[4] *Jian-yi*, "the change," the imperishable *Purusha*, *Brahma saguna*.
[5] *Bu-yi*, "the unchangeable change." In the *Bhagavad-Gîtâ*, explained Guénon, *Purushottama*, identical to *Paramâtmâ*, is beyond the two *Purushas* (perishable and imperishable), since it is the supreme Principle (that is, it is *Brahma nirguna* or the Non-Being) (*Symboles fondamentaux*, p. 451).

chrysolyte; the eighth, beryl; the ninth, a topaz; the tenth, a chrysoprasus; the eleventh, a jacinth; the twelfth, an amethyst.[1] And the twelve gates were twelve pearls: every several gate was of one pearl: and the street of the city was pure gold, as it were transparent glass.[2] And he shewed me a pure river of water of life, clear as crystal, proceeding out of the throne of God and of the Lamb. In the midst of the street of it, and on either side of the river, was there the tree of life, which bare twelve manners of fruits, and yielded her fruit every month: and the leaves of the tree were for the healing of the nations."[3]

The divine *Vaikuntha* and Heavenly Jerusalem are not temples built to symbolize the Center, but they represent the Center manifested as temple. And we should be aware that Heavenly Jerusalem is a precious stone: "that great city, the holy Jerusalem, descending out of heaven from God, having the glory of God: and her light was like unto a stone most precious, even like a jasper stone, clear as crystal"[4]; and a cube: "The length and the breadth and the height of it are equal."[5] René Guénon described the "solidification" of the present *Manvantara*

---

[1] The precious stones, insistently mentioned, refer to the celestial nature of these minerals. It is interesting that the breastplate of the Jewish High Priest (covering and marking the "Heart of the World") is equivalent to Heavenly Jerusalem: "And thou shalt make the breastplate of judgment with cunning work; after the work of the ephod thou shalt make it; of gold, of blue, and of purple, and of scarlet, and of fine twined linen, shalt thou make it. Foursquare it shall be being doubled; a span shall be the length thereof, and a span shall be the breadth thereof. And thou shalt set in it settings of stones, even four rows of stones: the first row shall be a sardius, a topaz, and a carbuncle: this shall be the first row. And the second row shall be an emerald, a sapphire, and a diamond. And the third row a ligure, an agate, and an amethyst. And the fourth row a beryl, and an onyx, and a jasper: they shall be set in gold in their inclosings" (*Exodus* 28:15-20).
[2] *Revelation* 21:10-21.
[3] *Revelation* 22:1-2. As Guénon observed, if the three squares of the Druidic precinct are rearranged, it is possible to obtain the image of Heavenly Jerusalem and of the twelve Zodiacal gates (*Symboles fondamentaux*, p. 104).
[4] *Revelation* 21:10-11.
[5] *Revelation* 21:16. The Holy of Holies in the Temple of Jerusalem was also a cube (Burckhardt, *Art of Islam*, p. 4).

as a gradual passage from sphere to cube, where the former is the shape of the Earthly Paradise and the latter of Heavenly Jerusalem.[1] Yet this change of shape has also a compensatory meaning: it must be viewed as a return to God's immutability, which is symbolized by the cube, Heavenly Jerusalem being, in comparison to the agonizing and decaying world (the "materialistic" and "hardened" cube of the end), the precious stone, the trustful mineral that does not change and does not perish. The great Ibn 'Arabî said that, among the natural reigns, the mineral is the one which possesses the greatest knowledge through Allâh and the greatest servitude to Allâh. It was created in Knowledge and it does not have intellect, wishes or intentions; it is always somebody else's wish or tendency to use a mineral; the mineral's intention is Allâh's intention. The vegetal reign, like the mineral one, is created in Knowledge, but it is on a lower degree because the plant grows and tries by itself to go higher (illustrating "arrogance"). The mineral is not like this, since it does not rise naturally; on the contrary, if it is raised and left free the mineral will try to descend, illustrating the essential reality of servitude. There is no other higher qualification in a man than the mineral qualification; after this comes the vegetal one, and then the animal one, because the animals are capable of moving in various directions (in comparison to the immutability of Kaaba). The lowest one is the man who thinks he is "god," when, in fact, at the end of the cycle he is completely decayed, profane, arrogant and without any notion of what real servitude means. The descent of the *Manvantara* affects man the most and the mineral the least.[2]

---

[1] *Le règne*, p. 191.
[2] Even the animals are more in accord with their nature than man is.

IV

THE TEMPLE AS A CENTER

It is said in the *Vastu Shastra*: "The best location to build a temple is at a *tîrtha* (*templum*, 'holy place')." As we already mentioned, a *tîrtha* is the isthmus or the ford between this world and the divine one (analog to the Islamic *barzakh*); crossing the *tîrtha*, the being ascends toward Liberation and the spiritual influences come into the world. Therefore a temple is a spiritual vehicle aimed at God.

René Guénon, in accord with Ananda K. Coomaraswamy, explained the symbolism of crossing a river and pointed out the spiritual significance of the *Pontifex*, identical to *Tîrthankara*, as well as the *Avatâra*, which refers to a "crossing" downwards.[1] A *tîrtha* being a "river ford," the appellative was extended to designate a "holy place" near a well, pond, lake, river or sea (with "holy waters")[2]; but also a sacred mountain or a place

---

[1] Guénon, *Symboles fondamentaux*, p. 344. Coomaraswamy specified that from *tr*, "to cross over," derived a great variety of terms such as *tara, tarana, târâ, tîrtha*, etc. *Tîrtha* is "crossing place," *tarana* is "crossing," hence *avatarana, tîrthakara* is virtually synonymous with *pontifex* (Ananda K. Coomaraswamy, *Selected Papers, Metaphysics*, Princeton, 1977, p. 325).

[2] We already mentioned the importance of the holy spring (the Castalian Spring and the spring of Chartres). Haridwâr, not far way from the Ganges' source, is also a "holy place." René Guénon mentioned the spring, corresponding to the primordial Ether, which in *Fedeli d'Amore*'s tradition is "the fountain of youth" (*fons juventutis*), always situated at the foot of a tree (Guénon, *Le symbolisme de la croix*, p. 67).

associated with a deity or a saint became a *tîrtha*.[1] With the evolvement of the cycle, *Purusha*'s sacrifice and division (multiplication) exacerbated and it became more and more difficult for the *templum* to acquire a direct divine blessing, which forced it to be content with the influence of a further centrifugal dispersed "part" of *Purusha*. At the beginning, God indicated directly the location of the "holy place"; later on, a place became holy and was considered a center due to a saint's influence or because a sacred event occurred there; eventually, when many temples as centers were built, the *necessary and sufficient condition* was to have holy relics included in the sanctuary. In Medieval Europe, when the Christian traditional society flourished, the holy places, the saints' tombs, and the relics proliferated considerably. As the Byzantine Empire had a dense sacred "web" of relics, so a similar "web" was created in the Occident,[2] often by using what is called *furtum sacrum*.[3] The holy "web" constituted, in fact, the spiritual infrastructure of the Christian world, since it was composed of sacred "knots" charged with spiritual influences, the relics being the "supports" of these influences.[4] René Guénon asked the rhetorical

---

[1] There is another name in the Hindu tradition for a "holy place": *kshêtra* ("field"); it is connected to *mandala* and it defines today a dwelling place of a saint or a temple.
[2] See our *The Wrath of Gods*, Rose-Cross Books, 2004, p. 169.
[3] It is well known how the Crusaders, the Venetians and others "stole" many relics from the Orient. The excuse was often that a particular saint wanted the relics to be brought to the place where supposedly the saint was buried. A legend describes how Saint James the Greater, though he suffered martyrdom in Jerusalem, was miraculously "moved" to Compostela. A similar "transfer" happened with the relics of the Three Magi: their tomb was in Hagia Sophia, at Constantinople, from where their relics (that is, three skulls) were moved to the Cathedral of Milan, and then, in 1164 (when Milan was conquered) they were brought to Cologne, when the city became one of the most important pilgrimage sites. Sometimes the relics arrived in the Occident carried by angels from the Orient: this is the case of the Holy Ladder (the ladder Jesus used when he climbed in front of Pilate) which, it is said, the angels brought from Jerusalem to Rome.
[4] We should mention that, in the Christian tradition, fountains were also considered holy places, visited by pilgrims, and sometimes a church was built near such a sacred fountain. We see the analogy with the *tîrtha*. The Temple of

question: "How could the Catholic worship of relics be justified, or the pilgrimage to the tombs of saints, if we do not admit that something non-material remains, in a way or another, attached to the corpse after death?" And he answered: "The forces we are talking about ... belong to a superior order ... We touch here the subject of 'spiritual influences.'"[1]

As the center is a contact point between Heaven and Earth,[2] so the temple is the mediator between God and men or, as Bede said, so is Christ Jesus the mediator between God and men, and Bede explained his affirmation: "He Himself attests when He says, *Destroy this temple, and in three days I shall raise it up.*"[3] Therefore, the temple is the junction point between the "center as a temple" and the "temple as a center"; in the first case, as Bede said, "He became the temple of God by assuming human nature," while in the second case, "we become the temple of God *through His Spirit dwelling in us.*"[4]

Solomon and the Ark of the Covenant were "condensers" of the spiritual influences (Guénon, *Aperçus sur l'initiation*, p. 167).
[1] René Guénon, *L'Erreur spirite*, Éd. Traditionnelles, 1984, pp. 58-59.
[2] Guénon, *Écrits*, p. 112.
[3] Bede 5.
[4] What we are saying here is possible only because the integral being combines the "divine" and the "human," which is a tremendous mystery. Ibn 'Arabî said: "The Garden of Paradise and this lower-world were combined in the adobe and in the building (of Adam-*insân*, the integral human being), even though one of them is made from clay and straw, and the other from gold and silver. This is an enormous question which he gave us as a riddle and then went away. Whoever knows its meaning will be happy and rested" (*The Reflective Heart*, p. 288). However, Ibn 'Arabî explained that the "temple as center" exists only to help us reach the "center as temple," or, in his words, our earthly existence should be used to reach the spiritual world: "the Fire of our earthly existence, with all its inescapable suffering and loss, is the ultimate attainment of human spiritual perfection," where the Fire alludes to the fact that the place of Gehenna is in our soul, as Coomaraswamy demonstrated later in his *Who is "Satan" and "Where Is Hell"*? See *The Reflective Heart*, pp. 284-292, 296. Similarly, the Christian tradition insists upon the human side as a mean to obtain salvation; St. Athanasius said: "The self-revealing of the Word is in every dimension – above, in creation; below, in the incarnation; in the depth, in Hades; in the breadth, throughout the world. ... For this reason he [Christ] did not offer the sacrifice on behalf of all immediately he came, for if he had surrendered his body to death and then raised it again at once he

The foundation of a terrestrial spiritual center is an act of God and in the "Golden Age" there was a direct and immediate projection of the Center as temple into the world.[1] This projection could be fathomed as a manifestation into the sensible domain, in the same way as the "earth" (*Dwîpa*) of a *Manvantara* manifested or emerged into the sensible domain.[2] This center was first situated at the Pole, and so the Primordial Tradition has a polar origin, and only in more recent epochs, when the seat of the Primordial Tradition was transferred to other regions, "it could become either Occidental or Oriental, Occidental for certain periods and Oriental for others, but surely lastly Oriental and well before the commencement of the 'historical' times."[3] Nonetheless, since all the centers are images of the Supreme Center, "the same appellations could have been applied, in the course of time, to very different regions, and not

---

would have ceased to be an object of our senses. Instead of that, he stayed in his body and let himself be seen in it, doing acts and giving signs which showed him to be not only man, but also God the Word"; Father Staniloae commented: "The Son of God does not scorn man's earthly life as worthless and meaningless; instead he adopts it as his own so that in the course of this very life the healing and sanctification of man may have its beginning. ... We ascend to the heavenly Christ through the earthly Christ. ... No path towards eschatological perfection exists which bypasses life on earth and the struggles which accompany that life" (Dumitru Staniloae, *Theology and the Church*, St. Vladimir's Seminary Press, 1980, pp. 193-194, 205, 207).

[1] We should not forget that Heavenly Jerusalem, for example, is the Center for our particular world and not for universal manifestation. The Earthly Paradise and Heavenly Jerusalem are identical (Guénon, *Symboles fondamentaux*, p. 333).

[2] Guénon, *Formes traditionnelles et cycles cosmiques*, p. 16. We note that the state of Existence to which belongs our human integral individuality contains not only the corporeal domain but also the domain of the subtle manifestation (Guénon, *Le symbolisme de la croix*, p. 74), and, consequently, the Center of the World is primarily the result of a manifestation of the Supreme Center into the domain of the subtle World; the Hindu tradition calls this manifestation of the Center *Hiranyagarbha*, the "Golden Embryo" (Guénon, *L'homme et son devenir*, p. 112).

[3] Guénon, *Formes traditionnelles et cycles cosmiques*, pp. 36-37. We saw that Seth, the elected one who was able to return to the Earthly Paradise (the Center) and recover the Holy Grail (a deed representing the restoration of the primordial order), could establish a spiritual center replacing the lost Paradise and being its image.

only to the successive locations of the primordial traditional center, but also to the secondary centers derived from it, more or less directly."[1]

At the same time with the "division" of the Primordial Tradition into different traditional forms, the divine projections multiplied and various "holy places" and secondary centers were born, all of them linked to the supreme Center. Only when the decadence became visible, and the bond with God weakened significantly, was the initiatory process introduced and the temple as center was built. Yet erecting a temple was not an arbitrary task, but it had to imitate the Act of God: first, a "holy place" had to be found, related to the ritual orientation[2];

---

[1] Guénon, *Formes traditionnelles et cycles cosmiques*, p. 38: René Guénon gave as an example Atlantis, and explained that the hyperborean Tula represented the first and supreme center for the entire present *Manvantara*; this center was the truest "sacred island," having literally a polar situation at the origin; all the other "sacred islands" were its images, which was the case of Atlantis. Tula of Atlantis had to be the seat of a spiritual power that emanated from the hyperborean Tula (p. 46); and the geographical situation of these centers is related to the doctrine of the cosmic cycles (p. 47).

[2] The orientation meant to find a "holy site," therefore a sanctuary was built over the vestiges of an ancient one; also, the temple had to be "oriented" in such a way that it could communicate with the Center, the same "orientation" being required by any neophyte following a spiritual quest: "we should talk of something that is hidden rather than truly lost, because it is not lost to everybody but some still possess it in its integrity; therefore, others have always the possibility to rediscover it, provided they search in a proper manner, which is to say that their intention must be directed in such a way that, through the harmonious vibrations it awakens, following the law of 'concordant actions and reactions,' it enables an effective spiritual communication to be established with the Supreme Center. In all traditional forms, this direction of intention is always symbolically represented; we are referring to the ritual orientation: this one, indeed, is properly the direction towards a spiritual center that – whatever it may be – is always an image of the true 'Center of the World'" (Guénon, *Le Roi*, p. 69). The building of a house needed also to respect the "sacred geography," even though each house was only an image of the center and could not each one be built in a "sacred place"; however, in India, for example, the *Brâhmanas* examined the soil, which was tasted, smelled, touched and looked at, in conformity with the four *tanmâtras* and the four *bhûtas* (the fifth, the Ether, represents the center). Moreover, the "sacred geography" had to be correlated with a "celestial geography" (Guénon, *Franc-Maçonnerie*, I, p. 55).

second, the temple had to be built obeying God's indications and following the plan of a divine archetype[1]; third, the temple had to be consecrated to becoming a "house of God"[2]; fourth, the temple was conceived not only as center, but as the image of the whole Cosmos, following the symbolic hierarchy of the spiritual Pole's successive approximations. This explains why the cyclic numbers were encrypted into the temple's proportions as an expression of time's spatialization,[3] since the sanctuary is not only, viewing its spatial form, an image of the Cosmos, but also, due to "solidifying" the laws of the cosmic cycles, the temple's measures and proportions organically harmonize with the truthfulness of the cosmic cycles (in this respect, the temple's orientation is essential for the accomplishment of harmony and perfection).

The mountain is one of the most prominent symbols for the Center. In various traditions, God's revelation is related to a sacred mountain,[4] since the shape and verticality of a mountain suggest the *Axis Mundi*, its peak alludes to "the Most High" and

---

[1] The Abbot Suger stated that the design of his church had been inspired by a celestial vision (Otto von Simson, *The Gothic Cathedral*, Harper Torchbooks, 1956, p. XVII); we should recall that Gothic art starts with Suger's St.-Denis. Today, France has abandoned St.-Denis, a holy place that was the spiritual and royal center of France.

[2] "Regarding the 'vivification,' we signify that the consecration of the temples, of the icons and of the ritual objects has as essential objective to make them effective receptacles of the spiritual influences, otherwise the rites to which they are designed to serve lacking any efficacy" (Guénon, *Aperçus sur l'initiation*, p. 59).

[3] We note that $3^2 \times 4^2 = 12^2 = 144$, and $9^2 \times 4^2 \times 10 = 12960$ (half of the equinoctial precession's duration). Wagner wrote in *Parsifal*: "Here the time changes into space," referring to the center of the world, Montsalvat (Guénon, *Le règne*, p. 215). However, we should recall that the Center of the World "resides, simultaneously, in the center of time and in the center of space" (Guénon, *Le symbolisme de la croix*, p. 31, *L'ésotérisme de Dante*, Gallimard, 1957, p. 66).

[4] René Guénon mentioned the mountain of the Terrestrial Paradise, the Hindu Mêru, Alborj of the Persians, Qâf of the Arabs, and Olympus of the Greeks (*Le Roi*, p. 73). "Upon the top of the mountain the whole limit thereof round about shall be most holy. Behold, this is the law of the house" (*Ezekiel* 43:12).

implies "inaccessibility,"[1] its solidity illustrates the immutability of the Principle, and the caves the mountain shelters within hint at the immanence of the Principle and at the hidden center. The mountain, Guénon said, epitomizes the Center of the World before the *Kali-yuga*,[2] and, to exemplify what was said above, we should add that, at the beginning of the present cycle, the mountain operated as a result of the supreme Center manifested as temple,[3] yet, in the *Kali-yuga*, the mountain became the model for the temple as center and the whole of India's architecture, for instance, supported by the plastic arts (sculpture and painting), had as objective to symbolize the center, either in the shape of a mountain,[4] or of a cavern.[5]

---

[1] The sanctuary is the "highest mountain" (Wensinck, *The Navel*, p. 13); "the mountain of the Lord's house shall be established in the top of the mountains, and shall be exalted above the hills" (*Isaiah* 2:2). The Paradise is in the top of a mountain higher than any mountain on earth, and therefore it was not attained by the waters of the deluge (Wensinck, *The Navel*, pp. 14-16). In the Judaic tradition, the Temple of Jerusalem is considered higher than the rest of the land of Israel and the land of Israel is higher than all other countries (*ibid.* p. 14); According to a *midrash*, the culminating point is the sacred rock (*The Labyrinth*, ed. S. H. Hooke, London, Society for Promoting Christian Knowledge, 1935, p. 54). In the Islamic tradition, it is said that "the highest of all countries and the nearest to Heaven is Jerusalem," but also that "In no place I ever saw Heaven nearer to earth than I saw it in Mecca" and "Kaaba is the highest situated territory, for it lies over against the center of Heaven" (Wensinck, *The Navel*, pp. 14-15).
[2] *Le Roi*, p. 74.
[3] Sometimes a sacred stone, a *beith-el*, substituted for the mountain.
[4] "The transition from the mountains to the sanctuary is a natural one" (Wensinck, *The Navel*, p. 11).
[5] "The apparition of the cave implies a displacement toward the interior, a covering of the center" (Benoist 39). For Isaiah, at the end of times, the center again will be at the top of the mountain: "it shall come to pass in the last days, that the mountain of the Lord's house shall be established in the top of the mountains, and shall be exalted above the hills; and all nations shall flow unto it ... And the idols he shall utterly abolish. And they shall go into the holes of the rocks, and into the caves of the earth, for fear of the Lord" (2:2, 2:18-19); "But in the last days it shall come to pass, that the mountain of the house of the Lord shall be established in the top of the mountains, and it shall be exalted above the hills; and people shall flow unto it" (*Micah* 4:1). "Ye are the light of the world. A city that is set on a hill cannot be hid" (*Matthew* 5:14).

This last phase, it is interesting to note, flourished in the time of Ashoka, when Buddhism was strong in India.¹ We observe the development of rupestrian architecture, probably based on ancient antecedents, since the mountain and the cave have been, as images of the center, the archaic sanctuaries in India.² The Buddhist temples (*chaitya*) or monasteries (*vihâra*) were built as caves by excavating the rocks, and later an inner *stûpa* represented the altar. The more and more opulent "ornamentation"³ exposed in a stronger explicit way the opposition between the immutable center and the perishable universal manifestation, illustrated by an indefinity of similar "ornamental" themes.⁴

In the Gupta Era, the temple visibly had the shape of a mountain, symbolizing the Center and the *Axis Mundi* supporting the multiple states of the being.⁵ Beside the sanctuaries excavated into the caves, there were temples carved and freed from enormous rocks,⁶ and we observe how the

---

¹ Guénon, *Introduction générale*, p. 176. See also René Guénon, *Autorité spirituelle et pouvoir temporal*, Didier et Richard, 1930, pp. 98-100; this text was suppressed in the later editions.

² "The Chinese art, more positive, starts from the idea of center and seed, an effective center, containing already in its mysterious complexity all the future developments" (Benoist 84).

³ We should underline that "ornament" means, in a traditional society, "a completion or fulfillment of the artefact" (Ananda K. Coomaraswamy, *Traditional Art and Symbolism*, Princeton, 1977, p. 242).

⁴ The "ornament," which usually was related to a deity, followed the same "explanatory" and descendant way from "complication" to "explication": *avyakta* (non-manifested; the images are symbols, like Shiva's *lingam*) – *vyaktâvyatka* (the images partially present a deity) – *vyakta* (manifested; the god is presented anthropomorphic).

⁵ "The altar and the temple, the throne and the palace, the city, the kingdom, the world refer to the same center, which is their model: Mount Meru" (Benoist, 80, 104). At Borobudur, between Heaven and Earth, there are eight states (Benoist 79); we note the number 8.

⁶ In a similar manner, the *apophatic* doctrine implies, after the example of the divine Dionysius the Areopagite, the removal from a marble block of those pieces that hide the statue. And Meister Eckhart said: "When a craftsman carves an image in wood or stone, he does not introduce the image into the wood, but removes the chips that hid and covered the image, he removes the slag; then shines what was hidden within." Ibn 'Arabî describes the spiritual

natural rock, mountain and tree are not replaced but transformed, in fact, explicated, into temples and columns, in order to illustrate, traditionally, the temple as center.

In the Khmer civilization,[1] the archetypal mountain theme was developed to extreme, the temples of Angkor,[2] comparable in many ways to the Egyptian pyramids and Aztec teocallis, representing famous applications. As was the Pharaoh of Egypt, who was king-priest – the Principle's projection into the world, who represented the synthesis of the whole Egyptian people (in this way his mummification and incorporation into the pyramid – divine vehicle – signifying the salvation of the entire Egypt), so was the Khmer king the saviour of his people, and the temple–mountain built while he was still alive symbolized the divine chariot that facilitated the king's Liberation (and of his people through him).[3] Of course, the absolute King is God: "For the Lord is our defense; and the Holy One of Israel is our

---

realization as a journey in God during which the composite nature is dissolved: "the spiritual traveler leaves behind in each world that part of himself which corresponds to it ... when he alone remains [without any of those other attachments to the world], then God removes from him the barrier of the veil and he remains with God (*The Meccan Revelations*, I, pp. 212-213); the same Ibn 'Arabî, when presenting the spiritual realization as an alchemical process, described a way consisting of the "elimination of evil" (which will change back the metal into gold): such an "elimination" takes place during the ascension; Jesus, Ibn 'Arabî affirmed, used this alchemical "procedure" when he healed the blinds and the lepers (Mohyiddin Ibn 'Arabî, *L'Alchimie du bonheur parfait*, Berg International, 1997, pp. 38, 39, 48, 67-68); we may note a German painting of the 18th Century (located in Heidelberg) where Jesus is presented as an alchemist.

[1] The Khmer Empire was a result of two powerful "Hinduist" kingdoms, called Fu-nan and Zhen-La by the Chinese; Fu-nan's king, who was also the supreme priest, bore the title of "king of the mountain." Even though a "Lord of the World," a king-priest, the Khmer king had to be consecrated by *Brâhmanas*, and he received its legitimacy from a *linga* erected at the top of a mountain, a *linga* that Shiva gave to the *Brâhmanas*.

[2] *Angkor* means "city," from Sanskrit *nagara*.

[3] As Dante was saying: "the purpose of the whole is to remove those who are living in this life from the state of wretchedness, and to lead them to the state of blessedness" (quoted in Coomaraswamy, *Christian and Oriental Philosophy of Art*, p. 107).

king,"[1] while the terrestrial king is his image, his progeny: "I have found David my servant; ... I will make him my firstborn, higher than the kings of the earth."[2] The Khmer sovereign, as an image of God, was the temple, and the temple was the Axial Mountain, the Center of the Universe.[3]

René Guénon quoted Mircea Eliade with respect to the mountain's symbolism: "In all these cultures (of the Semitic Orient, India and China), we encounter, on the one hand, the conception of the central mountain, which links various cosmic regions; on the other hand, the assimilation of a city, a temple, or a palace with this 'cosmic mountain' or their transformation, through the magic of the rite, into a 'center.'[4] Moreover, consecrating a space means, ultimately, to transform it into a 'center.' ... This process of identifying the houses, temples, palaces, cities with the 'Center of the World' is a spiritual phenomenon ... All the temples, all the cities, and even all the houses, regardless how far away they are in the profane space, dwell however in the same cosmic 'Center.'"[5] Nevertheless, what Eliade said must be regarded with caution. Without any doubt, in a traditional society, the Center, identical with the Principle and the Tradition, played a crucial role, and for a

---

[1] *Psalms* 89:18.
[2] *Psalms* 89:20, 89:27.
[3] For this reason, the sacred constructions proliferated and the whole population participated with enthusiasm in erecting the liberating temples; we could wonder if in Egypt the building of the pyramids was a punishment and a slave labour, or, in fact, had a similar motivation as in the case of the Khmer temple. The Khmer fervour should be compared to the passion of the Christian builders of the medieval cathedrals (the Christian temple was too erected not by force but by spiritual belief, see Burckhardt, *Chartres*, pp. 61-64: "Noblemen and noblewomen rush eagerly to seize the rope with which the heavy loads are pulled, and to place it on their shoulders, as if with this rope the land of promise itself were delivered unto them!"; the building of Chartres Cathedral "kindled in the entire population an enthusiasm unheard of until that time," Simson 148); there is an interesting similitude between the Khmer cycle and the Christian traditional society cycle.
[4] We have here what we call "the temple as a center."
[5] Guénon, *Comptes rendus*, pp. 188-9. See also Mircea Eliade, *Traité d'histoire des religions*, Payot, 1979, pp. 316-319.

society their own spiritual center represented the Center, the only true center, as their tradition was the only true tradition, but even in this situation, a house, a palace or a temple, different from the Temple or the Palace marking the secondary center, were recognized as being not "the same cosmic Center" but "images" of the center, connected to the real center, which, in its turn, had to be in communication with the supreme Center.

As we have seen, René Guénon is peremptory about this aspect. A secondary center is an emanation or a reflection of the supreme spiritual Center, which is the Center of the Primordial Tradition[1]; all the centers that are images of the veritable Center display common characteristics,[2] and their description is similar since they are emanations of the same supreme Center.[3] The secondary centers are virtually[4] identical to the supreme Center,[5] and in this sense it is possible to consider a house, a temple, a palace virtually identical to its particular spiritual center, but even if it is true that, in a traditional society, the art of constructing an edifice had, at the same time, an initiatory and a cosmogonic meaning, the unification of that particular edifice or sanctuary with the secondary center and, then, with the supreme Center, could happen only after the accomplishment of a spiritual realization. In this respect, the

---

[1] *La Grande Triade*, p. 138. The same thing Guénon said in *Écrits pour Regnabit*, p. 90.
[2] Guénon, *Le Roi*, p. 37. And again: "The secondary centers, due to the adaptation of the primordial Tradition to determined conditions, are, as we already said, images of the supreme center. The Sion could be in fact only such a secondary center, and though symbolically identical to the supreme center due to this similitude" (*Le Roi*, p. 57); and: "it is possible to have simultaneously, beside the principal center, several other centers attached to it and being its images" (*Le Roi*, p. 88).
[3] Guénon, *Le Roi*, p. 39.
[4] "Virtually" here means "not real."
[5] Guénon, *La Grande Triade*, p. 138.

more sanctuaries that were built, the less realistic the virtual identification with the supreme Center became.[1]

Mircea Eliade wrote: "The center of the world can be constructed *anywhere*, because one can build anywhere a microcosm in stone or brick"[2]; and: "the 'center of the world' could be ritually consecrated in infinite geographical points without the authenticity of each to jeopardize that of the others."[3] Eliade's opinion is polluted by the modern mentality, which has lost any notion related to "sacred geography" and the art of "orientation," especially in the Occident, where a sanctuary is built indeed *anywhere*.[4] Eliade follows Pascal who said: the space is "a sphere with the center everywhere" (about this see the first chapter of our present work). In fact, the indefinity of points is only "modalities" of the invisible supreme Point. Guénon said: "Each point ... is the potential center; yet, it must become effectively the center, by a real identification [that is, by "realization"], in order to make possible the total accomplishment of the being."[5] For Titus Burckhardt, "by the process of orientation, the sacred building is placed in the center of the visible universe. ... Of course, space, in its

---

[1] Guénon said: "The initiatory organizations that are farthest away from the supreme Center are also those which lose first the consciousness of the attachment to the Center" (Guénon, *Aperçus sur l'initiation*, p. 67).
[2] Mircea Eliade, *Symbolism, the Sacred, the Arts*, Crossroad, 1990, p. 138.
[3] Eliade, *Traité d'histoire des religions*, p. 200.
[4] René Guénon said: "it is not arbitrarily, indeed, that a specific place was chosen as center by organizations like the ones we discuss" (*Études sur la Franc-Maçonnerie et la Compagnonnage*, Éd. Traditionnelles, 1980, I, p. 9); see also what Guénon affirmed about the geographical situation regarding the places of pilgrimage, in connection to the sacred geography (*Franc-Maçonnerie*, I, p. 55). However, there are special cases, like those regarding a nomadic people (as the Jews of Moses, with their travelling Temple) or the case of the Templars (when they campaigned, they had a travelling church). See Gulielmus Durantis, *The Symbolism of Churches and Church Ornaments*, AMS Press, 1973, p. 199, about the "travelling Altar" used, with the bishop's permission, in the case of a journey. René Guénon stressed that, for a nomadic people, the spiritual center does not have to be a fixed place, and he gave the example of the Tabernacle (*Écrits*, p. 111).
[5] *L'ésotérisme de Dante*, p. 66.

immeasurable extension, has its center everywhere[1]; ... but sacred architecture gives this ubiquity of the spatial center its spiritual meaning, in that it firmly recalls that the true center of the world is here, where God Himself is present, in the sacrifice of the altar, and in the heart of the worshipper."[2] Once more, we should be cautious, since we must not imagine that being a Christian or a Muslim is enough to obtain salvation, nor, similarly, is it enough to go into a church or a mosque in order to reach the center.

Today, the temple as center becomes a "house of God" after its dedication,[3] and a home has to be blessed, but usually there are no more requirements related to the place where the edifice is built, and the blessing of a house has little to do with the symbolism of the center. Even in a traditional society, living in the *Kali-yuga*, the identification of a temple (house, palace, etc.) with the center, as its image, became, with the decadence of the cycle, more and more incomprehensible, and the crossing from virtual to real almost impossible.[4] In Eliade's opinion, there were two different "traditions": one regarded the initiatory way (the "difficult way") for reaching the center; the other one, regarded the desire of man to reach the Center of the World

---

[1] We see again Pascal's affirmation; Guénon said that, since the central point is essentially non-located (it has no location in manifestation), the *principial* center is, in fact, nowhere, and the circumference is everywhere (*Le symbolisme de la croix*, pp. 150-151).

[2] *Chartres*, p. 16.

[3] Durantis said that consecrating a church is done by divine prayers and holy unctions (p. 112); the dedication rites will bring *Shekinah* to dwell in the new built "house of God" ("to bring in a blessing," Durantis said, p. 118).

[4] Nevertheless, it seems that Eliade was just confused, since on other occasions he followed the right path. He admitted the need of a "holy place," validated by hierophanies, and transmitted over the ages from one people to another; the "place" was never chosen by man, but "discovered" with the help of *orientatio*, or indicated by a hierophany (an animal–guide, for instance); or it could be a place where a saint lived. Eventually, Eliade affirmed that "it is a 'center' any consecrated place where it could occur hierophanies and theophanies," but he thinks that to be real a new house or city has to be projected, by the construction ritual, into the "Center of the Universe," when the truth is that the city is a projection of the center (*Traité*, pp. 310-312, 315).

without any effort, and in this case any house was considered built in the very center.[1] There are not such "traditions" and Eliade is driven by sentimentalism when he talks about the man's "desire" to be "without any effort" in the center, the same sentimentalism he used when introducing melodramatic expressions like "the nostalgia for Paradise" and "the terror of history." What we have here in fact is the story of the temple as a center: the temple is an initiatory vehicle or a redeeming one,[2] which helps the neophyte to realize the temple within, that is, the center. It is a sacred place, which facilitates the communication with the superior domains; it is the image of the center, having as archetype the supreme Center; instead of a sanctuary, it could be a city,[3] a palace, or a house, having the same significance, yet if the inhabitants are not unified into the unique citizen these will remain "images" of the center, and nobody was so naïve as to think that the center would be reached without effort. The traditional construction of an edifice was done with effort, and in the same way as no one took the object symbolizing the deity to be the deity itself, no

---

[1] Eliade, *Traité*, p. 322.

[2] "From the vestibule to the sanctuary, the church illustrates and makes the catechumen advance on the road of redemption" (Benoist 64); "Forecourt, nave and transept corresponded respectively to the three stages of the Christian way spoken by the Fathers of the Church, namely purification, illumination, and union with God" (Burckhardt, *Chartres*, p. 13). The church itself symbolized the journey to the Center. "The church made by human hands, which is its symbolic copy because of the variety of divine things which are in it has been given to us for our guidance toward the highest good" (Maximus Confessor, *Selected Writings*, Paulist Press, 1985, p. 195).

[3] Suger calls the Gothic choir of the church "the City of the Great King" (Simson 130); on the other hand, for Suger the church is Solomon's Temple, and he does not have here in mind mere allegories, but archetypal realities, perfectly understandable for the medieval mentality and completely ignored by the modern one (Simson 133-134). As Coomaraswamy said, "If the medieval artist's constructions corresponded to a certain way of thinking, it is certain that we cannot understand them except to the extent that we can identify ourselves with this way of thinking ... medieval art was not like ours 'free' to ignore truth. For them, *Ars sine scientia nihil*: by 'science,' we mean of course, the reference of all particulars to unifying principles, not the 'laws' of statistical prediction" (*Christian and Oriental Philosophy of Art*, p. 29).

one thought the wheel to symbolize the physical sun, and no one assumed a house to be the very center.

Eliade should have recalled the text belonging to the traditional vestiges[1] of Eastern Europe he himself quoted and commented: "Down the (river) Argeş, through a lovely valley, comes the Black Prince with ten companions – nine great masters, apprentices, and masons, and the tenth, Master Manole, who surpasses them all. They are going together to choose a site for a monastery worthy to be remembered." And they ask a shepherd: "have you never, as you passed, seen an abandoned wall, left unfinished among pillars and hazels?" The shepherd shows them the way and they find the place. The Black Prince exclaims: "There it stands, my wall. Here I have chosen the place for my monastery."[2] The Black Prince represents the "spiritual architect," Master Manole the "royal architect," and the nine great masters are fellowcrafts and apprentices, where the number 9 refers, among others, to the circumference, while Master Manole symbolizes the center (the tenth), as projection of the celestial pole (the Black Prince).[3] Without doubt, the monastery could not be built anywhere, but in a "holy place."[4]

---

[1] About the traditional character of "folklore" see Coomaraswamy, *Christian and Oriental Philosophy of Art*, p. 138-140.

[2] Mircea Eliade, *Zalmoxis. The Vanishing God*, Univ. of Chicago Press, 1972, p. 166. About Master Manole and the Black Prince as Lord of the World see our works *Agarttha, the Invisible Center*, Rose-Cross Books, 2002, pp. 49, 33, 52 and *René Guénon et le Centre du Monde*, pp. 90, 99.

[3] A parallel could be made with the Christian tradition: the quest for the old wall (which marks a center) is similar to the quest for the newborn Jesus. We notice that not only are the Magi embarked on this quest (*Matthew* 2), but also the shepherds, who are guided by an angel (*Luke* 2:9-16), and in this second case the center is clearly mentioned: "Glory to God in the highest, and on earth peace, good will toward men" (see Guénon, *Le Roi*, p. 23). At the dedication of the church, the bishop entering the temple says: "Peace be to this house and to all them that dwell therein" (Durantis 121): an allusion to the temple as center. A *hadîth* says: You are the submissive and I am the Peace. My House is the House of Peace" (Ibn Arabi, *La niche des lumières*, Les Éditions de l'Oeuvre, 1983, p. 88).

[4] The same is valid for the foundation of a city. In one of Catherine Emmerich's visions she specified how the cities were built in special places,

Similarly, the Chartres Cathedral was built, the legend says, over the ruin of a druidic temple[1]; moreover, the Master Mason of the Gothic cathedral took great caution not to demolish the old, Romanesque church, and so did Suger with the Abbey of St.-Denis.[2] The tradition was kept unbroken; at Chartres, it is said, the cult of the Virgin has been practiced since time immemorial.[3]

The sacred art of medieval Europe was based upon the same universal principles as those, for example, existing in the Hindu

---

where three men similar to Enoch, long before Abraham, planted stones as landmarks (*Le Voile d'Isis*, 1932, no. 154, p. 629).

[1] Chartres was clearly the "temple as center": it was conceived as "the house of God and the gate of Heaven" (Simson 155), and a substitute center, since Suger chose it to preach the Crusade in 1150 (Simson 173).

[2] Simson 186. We should mention that when the cathedrals started to be built the Masons were ready to build them, since they belonged to an initiatory organization where the Art was transmitted uninterrupted from time immemorial; see for example the construction of the Cathedral of Chartres, in 1194, where the craftsmen (stonecutters, masons, carpenters, roofers, glass painters, etc.) were ready to operate (we should note the anonymity of these craftsmen; "the anonymity of the artist belongs to a type of culture dominated by the longing to be liberated from oneself"; "works of traditional art, whether Christian, Oriental or folk art, are hardly ever signed," Coomaraswamy, *Christian and Oriental Philosophy of Art*, pp. 39-41, 112, see also Guénon, *Le règne*, *Le double sens de l'anonymat*). Simson admitted that they possessed "perfect craftsmanship," and: "the building skills even of the early Middle Ages were far more highly developed than was long believed" (pp. 5-6), but this belief was part of the "great dupery," which started in the 17th Century and culminated in the 19th Century. Suger, for example, was aware that the Mason has to realize a spiritual harmony in his soul first to be able to build an edifice symbolizing this spiritual harmony (Simson 127), which is in accord with the initiatory meaning of the métier; Suger, like St. Augustine, considered that the Masonic process is "not so much the physical labour as it is the gradual 'edification' of those who take part in the building, the illumination of their souls by the vision of the divine harmony that is then reflected in the material work of art" (Simson 129) (we should also note the similarity between the thinking of Suger and of Maximus the Confessor, who was a close follower of Dionysius the Areopagite).

[3] Simson 219. Chartres was "the palace of the Blessed Virgin" (Burckhardt, *Chartres*, p. 83) and her "special chamber" (Simson 184).

tradition.[1] The Christian temple is, as any temple, an image of the center[2] and subsidiary of the Cosmos,[3] with Heavenly Jerusalem as main and divine archetype, but also carrying on the Judaic tradition with its Temple. For Gregory the Great, the city of Jerusalem and Ezekiel's Temple are both the Heavenly Jerusalem and the Church "which labours on earth before it will reign in heaven."[4] Bede, following St. Gregory, underlined that a church built by human hands is only an image, a copy of Jerusalem and its Temple, while the true building of the Church is a spiritual journey to the New Jerusalem.[5] For Bede, there is only one Temple: the spiritual one, and so it is for St. Bernard, who asked the Templars to think firstly about the "Greater Holy War," which takes place within us; and let us add that for a Mason the construction of the physical temple is just a "support" for the spiritual realization.

The Masonic initiation, with respect to Operative Masonry, had as support the construction of the Christian temple, which meant building the temple as center, yet with two implications: first, the Masonic act imitated God's creation of the Cosmos; second, from a microcosmic viewpoint, the initiation, followed by an effective spiritual realization, united the Mason with the

---

[1] Burckhardt used the syntagm "sacred art" too, but it is worthy to note that we need this expression only because of the present modern mentality, since normally in a traditional society there was no need to declare the art "sacred."

[2] "The holy Church of God is an image of God because it realizes the same union of the faithful with God" (Maximus 187).

[3] See Hani, *Le Symbolisme du Temple Chrétien*, pp. 33 ff. As Simson said about the symbolism of the cathedral, this one is "at once a 'model' of the cosmos and an image of the Celestial City. If the architect designed his sanctuary according to the laws of harmonious proportion, he did not only imitate the order of the visible world, but conveyed an intimation, inasmuch as that is possible to man, of the perfection of the world to come" (Simson 37). "On a second level of contemplation he used to speak of God's holy Church as a figure and image of the entire world composed of visible and invisible essences" (Maximus 188); Maximus the Confessor continues and considers the Church an image of man, and also of the soul (pp. 189-190).

[4] Bede XXVII.

[5] Bede LI.

supreme Center, when the temple as center could be realized within, in the heart.[1]

After the "holy place" was found, the orientation ritual followed, as in the case of the Hindu temple.[2] The orientation of the Christian temple, Burckhardt said, reflects the time or the cycle, expressed as space, while the Muslim mosque is orientated toward Mecca, toward a terrestrial center.[3] St. Isidore affirmed that a place facing east was called *templum* from *contemplating*, and when a temple was built, the architect took his East at the equinox,[4] and therefore, the next step in building the Christian temple is the foundation, which was "so contrived, that the Head of the Church may point due East: that is, to that point of the heaven, wherein the sun ariseth at the equinoxes."[5] As in the initiatory process, the spiritual influence, that is, the "non-human" element, is carried by rites, so in the building of a temple the rites are essential. In the case of the Christian temple, the spiritual authority (usually a bishop) sprinkles the place with holy water, to banish the evil forces,[6] and lays the first stone.[7] This first stone is an illustration of the words: "and

---

[1] A temple, but also any sacred and traditional edifice, has a "cosmic" significance, susceptible to a double application, macrocosmic and microcosmic (Guénon, *Symboles fondamentaux*, p. 261).

[2] See Hani, *Le Symbolisme du Temple Chrétien*, pp. 34-35.

[3] Titus Burckhardt, *L'art de l'Islam*, Sindbad, 1985, pp. 58-9. A particular mosque is comparable to a segment of a circle with Mecca as center. In the Christian temple case, the center is Christ. When the Muslims took possession of a church, first of all they destroyed the East-West axis and marked the direction to Mecca, the change in orientation symbolizing the change of the traditional form.

[4] Durantis 216.

[5] Durantis 21. Durantis was bishop of Mende (1237-1296), north of Montpellier. We may note how he complained about the priests' ignorance with regard to the sacred symbolism of the Christian ritual and church (Durantis 5), and this complain should be connected to the decadence of the Christian traditional society and the imminent end of the Templars.

[6] The sprinkled holy water is an image of the spiritual influence; there is a similitude with the rite of baptism.

[7] Durantis 21. "This stone must be placed at the North-East angle of the edifice" (Guénon, *Symboles fondamentaux*, p. 279). Suger wrote about the Abbey of St.-Denis' foundation: "Our noble king himself stepped down and laid a

the rain descended, and the floods came, and the winds blew, and beat upon that house; and it fell not: for it was founded upon a rock."[1] Regardless of how the symbolism of the stone was used in various traditional forms, its intimate meaning is related to the Center, since a stone, a rock or a mountain signify God and the Center: "Unto thee will I cry, O Lord my rock"[2]; "Forasmuch as thou sawest that the stone was cut out of the mountain without hands ...,"[3] which means that the first stone is an image of the center.[4] For Bede, the material temple was a figure of Christ "as the uniquely chosen and precious corner–stone laid in the foundation, and of us as the living stones built upon the foundation of the apostles and prophets, i.e. on the Lord himself."[5]

---

stone with his own hands; thereupon we ourselves and many other abbots and ecclesiastics laid stones. Many also included precious stones out of love and veneration for Jesus Christ, and sang: 'Precious stones are all Thy walls'" (Burckhardt, *Chartres*, p. 51).

[1] *Matthew* 7:25.

[2] *Psalms* 28:1.

[3] *Daniel* 2:45. Here the "stone" was considered a symbol of Christ and the mountain (or the rock) represented God, the supreme Center.

[4] Of course, all the other meanings remain valid, like this one: "And I say also unto thee, That thou art Peter, and upon this rock I will build my church" (*Matthew* 16:18).

[5] Bede 5-6. For Durantis, the "corner–stones," "which are of larger size, and polished, or squared, and placed on the outside and at the angles of the building, are men of holier life than others" (Durantis 22); "all the stones are polished and squared, – that is, holy and pure, and are built by the hands of the Great Architect into an abiding place in the Church" (Durantis 23). The symbol of the polished and squared stone is of great importance in Masonry: the passing from the rough ashlar (rude, unpolished stone) to the perfect ashlar (the squared and polished stone) symbolizes, as René Guénon said, the passing from "chaos" to the perfection and achievement of the "work" (*Symboles fondamentaux*, p. 316); it represents the spiritual realization. For Bede, the priests, similar to the stonemasons and woodcutters, try to change the ignorant "from the baseness and deformity in which they were born" and transfer him from the mountain of pride (from where the rough stone comes) to the mountain of the house of the Lord (built with squared and polished stones) (Bede 11-12, 15).

Ananda K. Coomaraswamy wrote an article to elucidate in what sense Christ was referred to as "corner–stone"[1] and René Guénon continued this investigation,[2] since there was confusion between the "keystone" or "angle–stone" and the "foundation–stone," confusion found not just in modern times, but already in "times when it is impossible to blame the ignorance of the symbolism."[3] Coomaraswamy and Guénon quoted St. Paul's words: "And are built upon the foundation of the apostles and prophets, Jesus Christ himself being the chief corner–stone; in whom all the building fitly framed together groweth unto an holy temple in the Lord: In whom ye also are builded together for an habitation of God through the Spirit,"[4] where "corner-stone," Coomaraswamy explained, should refer to a stone at the corner of a building, but since there are four such stones we cannot speak of the "corner-stone," which means that Christ is the "angle–stone" or the "keystone" – "the unique Principle upon which the whole edifice of the Church depends."[5]

This misunderstanding, Guénon said, would have meant to confuse St. Peter (the "foundation–stone") with Christ (the "angle-stone"), and he advanced the idea that it was, in fact, an intentional "substitution." What we have here is, of course, a consequence of the center's symbolism, its hierarchy and richness, and the confusion is similar to Eliade's confusion, because when the bishop lays the first stone, this one – the "foundation–stone" – is the only material element of the new church and comparable to a *beith-el* or an *omphalos*, representing the center, while, in reality, the "foundation–stone" is the *image* of the center: "we could say that each of the four 'corner–stones' reflects the dominant principle [Christ, the Center] of

---

[1] Coomaraswamy, *What is Civilisation?*, the article entitled *Eckstein* (1939).
[2] Guénon, *Symboles fondamentaux*, p. 278 (the article entitled *La 'Pierre angulaire*," 1940). This subject, Guénon specified, is closely related to the rituals of the *Royal Arch Masonry* (ibid. p. 280).
[3] Guénon, *Symboles fondamentaux*, pp. 278-279.
[4] *Ephesians* 2:20-22.
[5] Coomaraswamy, *What is Civilisation?*, pp. 168-169. The "corner" is also angular and hence the confusion.

the edifice, but it is not in anyway the principle itself."[1] Something analogous occurs in the case of the particular traditional forms: each one is an image, a reflection, an adaptation of the Primordial Tradition, and their centers are images of the Center, but somehow their members reach the belief that the particular tradition to which they belong is the only tradition and identical to the Primordial Tradition.

On the first stone, Durantis specified, "a cross must be engraved," which even more implies its role as image of the center; in the Eastern Church, a cross was placed where the church was to be,[2] but beyond its obvious meaning we should keep in mind that Guénon epitomized the Universal Man by a cross and he said: the Man, situated between Heaven and Earth, "not only belongs to the manifestation, but symbolically is its center."[3] For the Christian tradition, Jesus Christ is the Universal Man and the cross upon which the church is built represents His body; Christ crucified into the temple's diagram is, of course, similar to *Purusha*,[4] both identical to the Universal Man, yet we must nor forget that Christ is the Center[5] and the cross is also a symbol of this Center, not only because it is Christ's sign, but by itself. In fact, sometimes, the Christian symbolism of the cross overshadowed the most essential element: the center of the cross, which, in a way, is not visible, and from time to time had to be recalled, like the Rose–Cross did when they had the center of the cross marked by a rose or even having five roses, one in the center and four flanked by

---

[1] Guénon, *Symboles fondamentaux*, p. 280. "The 'first stone' or the 'foundation–stone' could be regarded as a reflection of the 'last stone,' which is the true 'angle–stone'" (Guénon, *Symboles fondamentaux*, p. 281); "the 'angle–stone' of the pinnacle is reflected in each of the 'foundation–stones' placed in the four corners of the base" (Guénon, *Symboles fondamentaux*, p. 284). In the Syriac literature it is said: "the center of the earth is situated higher than the four quarters" (Wensinck, *The Navel*, p. 14).
[2] Durantis 21.
[3] Guénon, *La Grande Triade*, p. 31.
[4] As Guénon said, any ritual, that is, a normal act in accord with order (Sanskrit *rita*), could be regarded as a sacrifice (*Symboles fondamentaux*, p. 304).
[5] "And I saw no temple therein: for the Lord God Almighty and the Lamb are the temple of it" (*Revelation* 21:22).

the cross' branches.¹ As René Guénon accentuated, the center of the cross is the "divine station" of the Islamic tradition, the Chinese *Zhong Yong* (the "Invariable Middle"),² where Cusa's *coincidentia oppositorum* is realized,³ and in this sense the cross is equivalent to a *swastika*, which is a sign of the Center, of the Pole, but also reveals the four "corner–stones."⁴ In the ancient rituals of the Operative Masonry, Guénon added, the letter G is represented in the center of the vault right above a *swastika* traced on the floor, symbolizing the connection between the celestial pole or center and the terrestrial one⁵; with respect to the Masonic enigma: "By letters four and science five,⁶ this G aright doth stand in a due art and proportion" and bearing in mind the central position of the letter G, Guénon introduced the equivalent Greek letter Γ, four letters Γ representing precisely the four branches of a *swastika*, but which, rearranged, could mark the four corners of a square, called in Arabic *el-arkân* (*rûkn* = angle, corner).⁷ This last diagram could be completed by adding the cross marking the center of the square, and René Guénon described it "as corresponding to a horizontal projection of an edifice: the four squares [*arkân*, Γ] correspond to the four foundation stones placed in the four corners [the "corner–stones"] and the cross to the "angle-stone" of the pinnacle."⁸

---

¹ Guénon, *Symboles fondamentaux*, p. 96, *Aperçus sur l'initiation*, p. 242.
² Guénon, *Le symbolisme de la croix*, pp. 49-50.
³ This expression, we recall, represented the Center's wall.
⁴ Guénon, *Le symbolisme de la croix*, pp. 70-72.
⁵ Guénon, *La Grande Triade*, p. 205.
⁶ Geometry (having the initial G) is the "fifth science."
⁷ Guénon, *Symboles fondamentaux*, pp. 139-140, 282-283.
⁸ Guénon, *Symboles fondamentaux*, p. 299. Guénon also quoted the description of the Heavenly Jerusalem given in the ritual of *The Chapter of the Sovereign Princes of the Rose Cross*, of *The Order of the Heredom of Kilwinning*, in which there is a golden square containing a cross and on the center of the cross lays a lamb (*L'ésotérisme de Dante*, p. 26); René Guénon underlined too that the circular enclosure of the Terrestrial Paradise encompasses a cross formed by the four rivers that sprang from the polar mountain (*Le Roi*, p. 93).

Sometimes, in the case of a dome,[1] the pinnacle is marked by an apical aperture, called the "eye of the dome,"[2] equivalent to *brahma-randhra*[3] and to the "strait gate" and to the "narrow way."[4] As Guénon noticed, with regard to the symbolism of the dome,[5] the point situated on the ground directly under the pinnacle of the dome is always virtually identical to the Center of the World, not as a topographic "location" but in a transcendent and *principial* sense, which means it can be "realized" in any consecrated and regularly established "center," hence the necessity of rites to ensure that the construction of an edifice is a veritable imitation of the making of the world. This point is, Guénon underlined, a true *omphalos*, and in many cases here the altar of the temple or the hearth of the house is placed.[6]

The architecture of the dome and vault flourished first in the Orient, from where it spread to Europe, as far back as in the

---

[1] "All the ritual architecture is related to the idea of center, either the symbolic center of the world or the spiritual center of man ... From an architectural point of view, the most perfect expression of this double symbolism would be the dome. ... All the circular constructions ... are images of Heaven" (Benoist 39-40).

[2] Coomaraswamy, *Traditional Art and Symbolism*, p. 441. This "eye" means "center."

[3] Guénon, *Symboles fondamentaux*, p. 269.

[4] Guénon, *Symboles fondamentaux*, pp. 272-3.

[5] He followed and developed Coomaraswamy's article *The Symbolism of the Dome* (*Traditional Art and Symbolism*, pp. 415 ff.).

[6] *Symboles fondamentaux*, p. 265. Usually the altar was made of stone, "because Christ signified by the Altar is the Stone growing into the mountain" (Durantis 152); yet Solomon made an altar of gold (*3 Kings* 7:48) and we know also that wood was the sacred material before stone: "And thou shalt make an altar of shittim wood, five cubits long, and five cubits broad; the altar shall be foursquare: and the height thereof shall be three cubits. And thou shalt make the horns of it upon the four corners thereof: his horns shall be of the same: and thou shalt overlay it with brass" (*Exodus* 27:1-2). Durantis mentioned an altar made of earth: "in the county of Provence, in the Castle of St. Mary by the Sea, there is also an Altar of earth, which Mary Magdalene, and Martha and Mary the mother of James, and Mary the mother of Salome, made there" (p. 153) (see also our work *The Wrath of Gods*, pp. 227 ff.).

times of the Roman Empire.¹ A famous example of a Roman dome is Agrippa's Pantheon built in Rome, which had the invisible center marked by the "eye of the dome," the apical aperture; but the dome was supported by a cylindrical base, while the Oriental dome had a cube as the base, and compared to the Latin dome, the Oriental one was built by accomplishing the passage from square to circle, or from cube to hemisphere.² Romanesque and Byzantine architecture embraced the Oriental dome and, at the same time, its symbolism,³ since one of the meanings of such a temple was that it really represented the Center where the two ends of the *Manvantara*, of our human cycle, unite, that is, the Terrestrial Paradise and Heavenly Jerusalem. The first, as Guénon said, corresponded to the beginning of the cycle and was circular, the latter corresponds to the end of the cycle and is squared,⁴ a symbolism reflected by the temple that has a circular roof (the Earthly Paradise) and a square floor (the Heavenly Jerusalem), which alludes to the fact that we are at the end of our cycle, since the Terrestrial Paradise is too far away, too high to be reached, while the Heavenly

---

[1] "As an architectural form, they [the rose windows] derived, like many other elements of the Gothic style – the pointed arch, the rib-vaulting, the tracery – from Islamic art." "Certain elements are already present in Romanesque architecture, but the decisive model derives from far way, namely from Islamic art, with which the Franks had for long been in touch" (Burckhardt, *Chartres*, pp. 46, 75). However, we should remember that the Arabs, like the Jews, were nomads and unfamiliar with the art of Masonry; they assimilated this art from the Byzantines (we mention among others the Armenians as very skillful builders) and the Persians ("Byzantine influence is present in the three great windows above the Royal Portal [at Chartres] ... There can be no doubt that the spell which the art of the Greek Church exerted upon the age of St. Bernard is also reflected in the first Gothic art," Simson 152). Also, Burckhardt considered that "one can discern a straight line of development from early Romanesque to Gothic" (*Chartres*, p. 29), yet René Guénon affirmed that "the passage from Romanesque to Gothic had to correspond to a change of conditions requiring a 'readaptation,' which would be realized in conformity to the traditional principles" (*Comptes rendus*, p. 45).
[2] To solve this architectural problem of the passage, two types of intermediaries were used: the pendentive and the trumpet arch.
[3] The Templars built their churches in the same way.
[4] *Le règne*, pp. 191, 218.

Jerusalem is the archetype of the Christian church.[1] However, there is also a well known "cosmic" symbolism, with the square symbolizing Earth and the dome itself Heaven,[2] which could be found also in the form of the Buddhist *stûpa* and Islamic *qubbah*, and Guénon stressed that in the initiatory ritual of the *Royal Arch Masonry*, the passage *from square to arch* (which is described as *exaltation* in the Royal Arch degree) means indeed a passage from Earth to Heaven.[3] From a cosmogonic viewpoint though, the passage is *from arch to square*, from unity to multiplicity, where the unity is symbolized by the "eye of the dome" or the "angle–stone," as Ananda K. Coomaraswamy stressed, commenting on St. Paul's text previously quoted[4]: "The evident intention of the text is to depict the Christ as the *unique* principle upon which the whole edifice of the Church depends. The principle of anything is neither one among parts of it, nor a totality of parts, but that in which all parts are reduced to a unity without composition."[5]

René Guénon added, after he described the dome as symbolizing Earth (square) and Heaven (circle): "to this general meaning, we can adjoin another one, more precise: the assemble of the edifice, regarded from top to bottom, represents the passage from the principial Unity (to which corresponds the

---

[1] The temple represents in this respect the realization of what the Hermeticists symbolically called the "quadrature of the circle" ("squaring the circle") (Guénon, *Le Roi*, p. 93). At the dedication of the church, the twelve crosses painted on the four walls and the twelve candles placed around the church (three at each cardinal point) (Durantis 125, 237) suggest the Heavenly Jerusalem and, of course, the Zodiac.

[2] Sometimes, there is no dome, which is replaced by the sky above (see Guénon, *Symboles fondamentaux*, p. 265); a famous example is the mosque Jama Masjid of New Delhi (built by Shah Jahan, who also was the builder of the Taj Mahal).

[3] Guénon, *Symboles fondamentaux*, pp. 261-2. "[Allâh] is the One Who has made the Earth a carpet for you and had the Sky built above you" (*Qur'an* 2:21); and a commentary added: "the building of Heaven over the Earth has the form of a cupola" (Wensinck, *The Navel*, p. 43).

[4] *Ephesians* 2:20-22.

[5] Coomaraswamy, *What is Civilisation?*, pp. 168-9, Guénon, *Symboles fondamentaux*, p. 279.

central point or the pinnacle of the dome) to the quaternary of the elementary manifestation[1]; from an architectural perspective this passage is realized (as we can see in the case of Hagia Sophia) by the pendentives.[2] Guénon resumed then the initiatory meaning, following Coomaraswamy: "the reverse process is represented by a Buddhist legend in which Buddha, having received four alms bowls from the *Mahârâjas* of the four cardinal points, made one single bowl, which indicates that, for the 'unified' being, the 'Grail' is again unique like it was at the beginning."[3] Only when a being realizes this "unification," the four "corner–stones" are reabsorbed into the "foundation–stone" *shethiyah* (placed in the center of the square) and this one is reabsorbed into the "angle–stone."

The essential difference between the center of the square and the "angle–stone" is the difference between *Square Masonry* and *Arch Masonry*, and only by "passing from square to compasses" is it possible to eliminate the difference between the secondary center and the Center.[4] In this Center, multiplicity returns to Unity, the parts of *Purusha* are unified, and, likewise, the being is reintegrated in the center of the cross marked by the rose. Dante described the Center as a Rose (*candida rosa*) where the multiplicity of saints and pure souls are

---

[1] Guénon, *Symboles fondamentaux*, p. 263. It is the passage from the "angle–stone" to the four "corner–stones."

[2] The four "corner–stones" correspond to the four horns of the altar. At the consecration of the altar, the bishop makes four crosses with the holy water at the four horns of the altar, and one in the middle; the middle cross signifies, Durantis said, "the Passion which Christ underwent in the middle of the earth, that is, in Jerusalem" (Durantis 144-145).

[3] Guénon, *Symboles fondamentaux*, pp. 263-264.

[4] It is interesting to mention that René Guénon considered the "keystone" or "angle–stone" a "stone descended from heaven" and an equivalent of the Grail (*Symboles fondamentaux*, pp. 286-287). It is obvious that the Holy Grail is a symbol for the Center, but also we see how the Center manifests itself as temple through the "angle–stone," which is "in whom all the building fitly framed together groweth unto a holy temple in the Lord." In addition, Guénon underscored that the "angle–stone" is not only a "stone descended from heaven", but also its position is "heavenly", since the base and the roof correspond to Earth and Heaven (*Symboles fondamentaux*, pp. 295-296).

fused (but not confused) into Unity,[1] and this Center is as well the spiritual Temple built with "living stones."[2]

Everything in the Christian temple reminds us of the Center, besides the fact that all the physical elements of a church and the rites illustrate the cosmogonic process as a replica of the spiritual Christic process: "He was crucified, died and was buried. On the third day He rose again."[3] The incense ritual underlines the cross inscribed in a circle marking the Center, but recalls too the initiatory and cosmogonic processes.[4] In a way, the church's bell is a symbol of the center as origin of *Parashabda*; and so it is the baptismal font, a *fons juventutis*, an "engendering womb" (as divine Dionysius said), *jîva-ghana* ("the reservoir of life"), *Malkuth*. Yet, we have to keep in mind that all these elements are just reflections of the center. The legend

---

[1] *The Divine Comedy, Paradiso*, XXX-XXXI. For Guénon, here Dante alluded to the white mantles of the Templars and to the Masonic convent ("mira quanto è il convento delle bianche stole!," *Paradiso*, XXX, 128-9) (*L'ésotérisme de Dante*, p. 25).

[2] Bede 5, Durantis 17 ("The material typifieth the spiritual Church"). Durantis said too: "the soul ... is the temple of the True God"; "the Catholic Church herself, made one out of many living stones, is the Temple of God, because many temples make one Temple, of which the True God is one" (Durantis 116). During the medieval pilgrimage to Compostela each pilgrim brought a stone, an image of the "living stone." The great spiritual master Ibn 'Arabî had a vision of Kaaba built of alternate gold and silver bricks, but though the building was complete, two bricks were missing: one gold and one silver; and Ibn 'Arabî saw himself replacing the two bricks and now Kaaba was perfect. Also in a *hadîth* the Prophet Muhammad compared prophethood to a wall of bricks, where he was the missing brick (Michel Chodkiewicz, *Seal of the Saints*, The Islamic Texts Society, 1993, pp. 123, 128). For Bede, the foundation stones are the prophets and apostles (Bede 14).

[3] Any initiatory ritual comprises a descent to hell, a death and a new birth, ascension to heaven (the ascendant realization) and a descent into the world (the descendant realization). For an *avatâra*, there is first the descent into the world and then follows the paradigmatic spiritual realization. For Durantis, the thrice-repeated circuit of the church, during the ritual of its dedication, represents: "the first was that by which He came down from Heaven to the World; the second in which He descended into Hell from the World; the third in which returning from Hell and rising again He ascended into Heaven" (p. 119) (see also Burckhardt, *Chartres*, p. 57).

[4] Hani, *Temple*, pp. 197 ff.

describes how Clovis, blinded by the luminosity of the baptismal font, asked St. Remy if this is the promised Kingdom of God, and the saint replied: "no, it is the entrance on the way heading to the Kingdom."[1] In fact, the temple itself is an "entrance," the "gate" to Heaven,[2] which makes the very gate of a temple so rich in symbolic meanings and almost a double of the temple itself.[3]

The center of the temple, that is, the center's center, the pure *omphalos*, is evidently the altar.[4] It represents the "kernel," the essence of the sacred edifice, Agni's hearth, the true *Beith-El*, the Holy of Holies, the stone *shethiyah* that sustains the Ark of the Covenant,[5] the dwelling place of *Shekinah*; it is Jacob's anointed stone and consecrated in the same way by the Christian priest, the paradigmatic gate of Heaven; it is Christ.[6]

---

[1] Patrick Demouy, *Notre-Dame de Reims*, CNRS Éd., 2001, p. 77.

[2] "This is none other but the house of God, and this is the gate of Heaven" (*Genesis* 28:17). See also the meaning of Babylon's name. However, the supreme center is a "temple without gates" (Guénon, *Aperçus sur l'ésotérisme islamique*, p. 125).

[3] In the Hindu tradition, the gates (*gopuram*) were sometimes more grandiose than the temple itself. "Then said Jesus unto them again, 'Verily, verily, I say unto you, I am the door of the sheep. ... I am the door: by me if any man enters in, he shall be saved, and shall go in and out, and find pasture'" (*John* 10:7, 9). For St. Symeon the New Theologian, "the Son is the gate, the Holy Spirit is the key, and the Father is the house." The gates of Chartres Cathedral represent one of the most complete expressions of the Christian doctrine (Burckhardt, *Chartres*, p. 65). Like in the case of the center, to pass through the gate you have to become the gate; on the door of Basel Cathedral a Master Mason was represented offering to Christ a gate-model; that is, he, as gate, offers himself to Christ–the Gate (Burckhardt, *Principes*, p. 117).

[4] The word "altar" derives from Latin *altus*, "high"; usually the altar stands on a three-stepped podium, crowning the Three Worlds. "Where is the navel? In Jerusalem. But the navel itself is the altar" (Wensinck, *The Navel*, p. 41).

[5] See Guénon, *Symboles fondamentaux*, p. 294.

[6] The Church Fathers agree in comparing the altar with Christ: "and that Rock was Christ" (*1 Corinthians* 10:4), St. Peter being just a "substitute." "To whom coming, as unto a living stone, disallowed indeed of men, but chosen of God, and precious, Ye also, as lively stones, are built up a spiritual house, a holy priesthood, to offer up spiritual sacrifices, acceptable to God by Jesus Christ. Wherefore also it is contained in the scripture, Behold, I lay in Sion a chief corner stone, elect, precious: and he that believeth on him shall not be

The altar is the Holy Heart of Christ, but also the Heart of the World, with which the heart of the one who accomplished the spiritual realization identifies.[1]

In the Orthodox Christian Church, the altar's leg must contain holy relics, and such relics have also to be sewed into the holy pall,[2] since they, together with the dedication of the church ritual, make the temple indeed an image of the center, a "holy place," and "house of God."[3] Similarly, Durantis said with regard to the Catholic Church: "Without the relics of Saints, or, where they cannot be had, without the Body of Christ, there is no consecration of a fixed Altar" (p. 149); and added: "between the two candlesticks the Cross is placed on the Altar: because Christ standeth in the Church, the Mediator between two peoples. For He is the corner–stone" (p. 71).

The cross marks the center of the center (the altar) of the center (the temple).[4] Yet, after the temple is built, another cross is traced during the dedication ritual: "A Cross made with ashes and sand is described athwart the church, upon which Cross of

---

confounded. Unto you therefore which believe he is precious: but unto them which be disobedient, the stone which the builders disallowed, the same is made the head of the corner, and a stone of stumbling, and a rock of offence, even to them which stumble at the word, being disobedient: whereunto also they were appointed" (*1 Peter* 2:4-8).

[1] "The Altar is our heart, on which we ought to offer"; "the Altar, ye know, sometimes signifieth the Heart" (Durantis 44, 47, 71). The altar marks the Center of the World, as Black Elk stressed when describing the Oglala Sioux's ritual for building the altar. We mention briefly that Burckhardt presented this description (*Principes*, p. 27), quoting Schuon's translation (Hehaka Sapa, *Les rites secrets des Indiens Sioux*, Payot, 1953), but at the indicated page 22 there is no such description; in fact, the altar ritual is recorded (in a slightly different way than the one presented by Burckhardt) at page 118 (see also Joseph Epes Brown, *The Sacred Pipe*, Univ. of Oklahoma Press, 1989, p. 89).

[2] The pall is called αντιμένσιον in Greek, "on the table," a liturgical object of the Byzantine rite. It is impossible to celebrate a holy liturgy without using holy relics.

[3] As Guénon explained, the relics and the rites are vehicles of the spiritual influences. Only after its anointment is a church transformed into what its name designates, Nicholas Cabasilas said.

[4] Guillaume Postel expressed the "center of the center" as "the Wheel in the center of the Wheel" (Guénon, *La Grande Triade*, p. 190).

dust the alphabet is written in the shape of a Cross in letters of Greek and Latin, but not of Hebrew ... and it is written with the Pastoral Staff."[1] The rite of writing with the Pastoral Staff is comparable to the Islamic doctrine regarding the supreme Pen (*al-Qalam al-a'la*) (identical to the primeval Intellect, *al-'Aql al-awwal*) used by Allâh to inscribe all the destinies on the preserved Tablet (*al-Lawh al-mahfûz*).[2] The Cross, which is a St. Andrew's cross, an X, is not only the initial of Christ's name and so its symbol, but also marks the center and reunites the four corners of the temple with the center.[3] It seems that, in the early years of Christianity, the Hebrew alphabet was inscribed too, the three alphabets representing at first sight the three languages of the Christian tradition and attesting to the fact that Christian European society was the heir of Greco-Roman and Judaic civilizations.[4]

As long as the Christian traditional civilization lasted, the cathedral, placed in the center of the city, was the "symbol of the kingdom of God on earth."[5] René Guénon considered the obliteration of the Order of the Temple, in 1312, the event that marked the end of the Christian traditional society,[6] when the

---

[1] Durantis 122. See also Burckhardt, *Chartres*, p. 57, Benoist 62, Hani, *Temple*, pp. 55 ff. For Durantis, the Pastoral Staff is the Divine Word (p. 120).
[2] See, for example, Titus Burckhardt, *Introduction aux doctrines ésotériques de l'Islam*, Derain, 1955, pp. 71-72. Inscribing the alphabet illustrates a cosmogonic process.
[3] In Masonry, the twenty-ninth degree of the Ancient and Accepted Scottish Rite is called *Grand Scottish–Knight of St. Andrew* (or *Patriarch of the Crusades*), having an X cross as jewel, and its legend connects the knights with the art of building Christian churches (Albert G. Mackey, *An Encyclopaedia of Freemasonry*, The Masonic History Company, 1924, I, p. 397).
[4] Sometimes, the alphabet was inscribed around the crown of the bell (Durantis 239). In Masonry, the ineffable Word is composed of elements belonging to Hebrew, Chaldean and Egyptian languages.
[5] Simson XVI.
[6] For René Guénon, the true Middle Ages lasted from the reign of Charlemagne to the beginning of the 14th Century, when a new decadence started and, following various stages, augmented to our days; this is the starting point of the modern crisis and of the desegregation of Christianity (*La crise*, p. 29).

links with the spiritual Center of the World were broken,[1] and when, as a consequence, the gradual transformation of the temple into a "lost center" started. "In Europe, every consciously established link with the center intermediated by regular[2] organizations is now broken, as has been the case for several centuries; however, this severance did not happen suddenly, but followed a series of successive phases. The first of these phases occurred at the beginning of the 14th Century; what we have already said about the Orders of Chivalry should make one understand that one of their main roles was to secure a communication between East and West, the importance of such a communication clearly resulting from the fact that the center we have in mind here was always described, at least with respect to the 'historical' times, as situated in the Orient."[3] The destruction of the Order of the Temple continued with the *Hundred Years War*, a conflict between France and England, lasting 116 years from 1337 to 1453, which exhausted most of the traditional elements still existing in Europe and prepared the way for the Renaissance and the Reformation.[4] "However, after the destruction of the Order of the Temple, the Rosicrucianism, or what later was called so, continued to ensure the same connection, even though in a more dissimulated manner. The Renaissance and the Reformation marked another critical phase, after which, as Saint-Yves appears to suggest, the complete and

---

[1] "Any conscious connection with the spiritual center of the world ended up broken, which means in a particular sense that the tradition was lost, mainly with regard to one or another secondary center, which ceased to be in direct and effective relation with the supreme center" (Guénon, *Le Roi*, p. 69).

[2] "Regular" – a word meaning "orthodox," "in conformity with the rules."

[3] Guénon, *Le Roi*, p. 70. As Guénon insisted, the Orders of Chivalry were, during the times of the Crusades, the main liaisons between Orient and Occident, which facilitated active intellectual exchanges; and we have to abandon the common image about exclusive hostile relations (Guénon, *L'ésotérisme de Dante*, p. 20).

[4] As Guénon said, the Renaissance and the Reformation represent not the beginning but the consequences of the decadence, which started almost two centuries earlier (*La crise*, pp. 29, 97).

final rupture coincided with the treaties of Westphalia, which in 1648 ended the Thirty Years War."[1]

In the 14th and 15th Centuries, the importance of the church as center started to be challenged by the town hall and the palace. The building of churches, the main goal of the traditional craftsmen, in which all the arts were employed, commenced to lose its significance.[2] The Renaissance[3] also facilitated the manufacturing of the "great dupery," which culminated in the 19th Century; the so called return to the art of the classical antiquity implied a shortcut of the medieval traditional times, which became the "dark ages."[4] The Reformation joined this subversive plan and did the same thing from a religious viewpoint, aggressively and foolishly promoting the picture of an "evil" Middle Ages, stationary and closed, which had to be erased from history to allow progress to take place.[5]

---

[1] Guénon, *Le Roi*, p. 71.
[2] "The religious sentiment had become pure aesthetic"; see Hans Sedlmayr, *Art in Crisis, The Lost Center*, Transaction Publishers, 2007, pp. 11-13 (It is worth noting that, in USA, the author could sell only 240 copies because he attacked the modern mentality).
[3] "What is called Renaissance was in fact the death of a lot of things; pretending to return to the Greek-Roman civilization, it took possession of what this civilization had the most superficial" (Guénon, *La crise*, p. 29).
[4] This is what Guénon called the "classic prejudgment" (that is, when the Westerners, starting with the Renaissance, decided that they are the inheritors of the Greco-Roman antiquity) (*Orient et Occident*, Guy Trédaniel, 1993, p. 28).
[5] See Anthony Kemp, *The Estrangement of the Past*, Oxford Univ. Press, 1991, pp. 98-100, 103-104. Petrarch, in a letter of 1359, considered the period between the Roman civilization and Renaissance an age of *tenebrae* (Kemp 98). As Kemp said, Renaissance did not precede the Reformation but was contemporaneous with it; also, the Renaissance was not a secular movement, nor was the Reformation antihumanistic; they were both alike in their critical methodology (it is just a part of the "great dupery" to envision the two as opposing each other). Regarding the Reformation, we should add René Guénon's firm conviction that there are not such things as spontaneous movements, either in the political order or the religious order; always it has to be an impulsion, which the apparent leaders of those movements could ignore its origin (Guénon, *L'Erreur spirite*, Éditions Traditionnelles, 1984, pp. 26). It seems that the Rosicrucians tried to inspire, without success, a "reformation" of the decaying Christian society; and Luther was a sort of a "subaltern agent"

Breaking the link with the spiritual center symbolized in fact breaking communication with Heaven and a "fall" of the eyes, which instead of looking upwards[1] started to look exclusively downwards; to this dramatic change belongs what is called "humanism," and which was credited to the Renaissance.[2] As we will see in the next chapters, human nature is plainly present in the sacred texts and the reader is amazed, if not shocked, to witness all types of "sins" taking place, which trigger God's punishment. The history of Judaism, Christianity or Islam, to name only the three religions, is imbued with wrongdoings,[3] and there is a reason why the Christian Church insisted upon the "human nature" of Christ, even though later it was misinterpreted. We have to remember that the three religions developed during the *Kali-yuga*, and in better times the sacerdotal function should have belonged to an individual initiated in the "Sacerdotal Art" and the royal function to an individual fully initiated in the "Royal Art"[4]; but, in the "Iron

---

(Guénon, *L'ésotérisme de Dante*, p. 25; about Luther's coat of arms containing the cross encircled by a rose, see Sédir, *Histoire et Doctrines des Rose-Croix*, Bibliothèque des Amitiés Spirituelles, 1932, p. 64). There are similarities between the Reformation and Spiritism (or Spiritualism), regarding their "evolutionist" and "pseudo-spiritual" views and their origins (we have in mind here both the forces that inspired them and those that took control of them); it was said that the H. B. of L. inspired the spiritist movement to fight the materialism (also it was said that the H. B. of L. was responsible for the setting up of the Theosophical Society, see Joscelyin Godwin, Christian Chanel, John P. Deveney, *The Hermetic Brotherhood of Luxor*, Samuel Weiser, 1995, p. 25) (see Guénon, *L'Erreur spirite*, pp. 24-29, 276-8, 404).

[1] The meaning of *anthropos*.

[2] See Guénon, *La crise*, p. 31. Guénon said: "It means to reduce everything to human proportions only, to ignore any superior principle and, we could say symbolically, to turn away from Heaven with the pretext of conquering the earth." For Simson, the Renaissance is characterized by "the self-centered and quarrelsome individualism" (p. 221).

[3] To give just one example, the Christian apostles acted utterly "human" when Peter denied Jesus, Judas betrayed him, and the apostles were afraid to watch the crucifixion.

[4] See Guénon, *Autorité spirituelle*, p. 36. Even the name "Sacerdotal Art" disappeared in Renaissance times, which marks, from all viewpoints, the consummation of the break between the Occidental world and its own traditional doctrines. Some established the year 1459 (immediately after the

Age," too often the individual did not match the function, and the secondary cycles multiplied and precipitated with increasing haste. Therefore, even though the Reformation wanted regeneration of humankind and of the Church, it was incapable of understanding the difference between divine and human, and, mistaking the function with the individual, became in many ways anti-traditional,[1] replacing spirituality with morality and humanism. The representatives of the Reformation completely ignored the fact that *perfection* is an attribute of the center and individual initiatives cannot build a "terrestrial paradise" governed by morality and other human elements, and likewise the "Golden Age" will come only with the "reversal of the poles."[2]

What seems extraordinary, René Guénon said, is "how rapidly the medieval civilization was completely forgotten; the people of the 17th Century had no notion at all about it, and the monuments that survived it did not represent a thing for them,

---

end of the *Hundred Years War*) as the date when the ancient tradition was lost and the guilds of builders had to reorganize themselves (*ibid.* p. 37).

[1] There are many cases today when people foolishly motivate their "atheism" by arguing that the individual who operates the sacerdotal function is unworthy. Frithjof Schuon introduced the expression *la marge humaine* (the human margin or limit) to describe how the divine influence is total only for the Scripture and the essential consequences of the Revelation, and there is always a "human margin" where this influence is indirect; the fallen or post-Edenic man is a quasi fragmented being and the sanctity of a man does not stop him for being sentimental or a weak logician; the most typical examples of "human margin" are those supplied by the separation of orthodox religions, Schuon said (Études Traditionnelles, 1970, no. 421-422).

[2] It is well known how the Protestant immigrants (and especially the Puritans) identified the New World with the "lost paradise"; see Mircea Eliade, *La nostalgie des origines*, Gallimard, 1971, Kemp 135; Cotton Mather in his *Magnalia Christi Americana* said: "In short, the *first Age* was the *golden Age*: to return unto that, will make a man a *Protestant*, and I may add, a *Puritan*" (Kemp 142). Kemp describes how the Reformation opened the door to the concept of continuous progress and revolution, each new "revolution" considering the previous protestant movement decayed and in need of reestablishment of the truth (there is no surprise that, when the idea of continuous revolution was imported by the communist regimes, they also adopted this phrase: "the purity of the members of the party," in accord with the Puritanism). See Kemp 109-110, 128-9, 134-5, 139, 142, 147-8, 150.

neither in the intellectual order, nor even in the esthetical one."[1] And Guénon insisted that it is difficult to accept such a change as spontaneous and not influenced by an enigmatic will. "It is strongly implausible also that the legend that presented the Middle Ages as a 'dark age,' of ignorance and barbaric, was born by itself."[2]

Starting with the 17th Century and continuing into the following centuries,[3] the temple becomes the "lost center" and it is replaced by the profane palace, by the English garden (the "paradise"),[4] by the Museum (where the anonymity of the artist is substituted by a ridiculous cult of the craftsman's individuality), by the Theatre or Opera (which is built in the center of the city),[5] by Exhibition, by the Railway Station

---

[1] *La crise*, p. 30.
[2] *La crise*, p. 31.
[3] It is not coincidence that at the end of the 18th Century the caricature (which was known, of course, in the previous centuries) is promoted as art and becomes widespread, illustrating the degradation of man (Sedlmayr 124-126). And we must notice how today cartoons (as well as computer games) are highly appreciated by adults, since we witness a process of infantilization; Sedlmayr said: "Hereabouts lie the roots of the love of the absurd and non-sensical that is a mark of modern man, a love that is exemplified in the comic cartoon films of the twenties and particularly in the work of Walt Disney, though now the irrational element has been translated into an idiom of innocence, a fairy idiom of pure fun" (p. 129). There are, of course, notable exceptions, they will always be, like William Blake, for example, who said in his *Milton*: "I will not cease from mental fight,/ Nor shall my Sword sleep in my hand,/ Till we have built Jerusalem/ In England's green and pleasant Land." We should also mention that it was in the 18th Century when the last *autos sacramentales* (the Christian mysteries) were suppressed in Spain (year 1765), as a triumph of the antitraditional forces; in France, under the pressure of the Reformation and atheism, the sacred theatre was prohibited already in 1548 (see René Allar, *Des "Autos Sacramentales"* in *Le Voile d'Isis*, no. 156, 1932). As Guénon said, in the middle of the 18th Century, the ideas like "civilization" and "progress," as well as the materialism, were born (Guénon, *Orient et Occident*, p. 24).
[4] We should mention here the floral labyrinth, which, like the earlier "turf mazes," suggested the center.
[5] Regarding the theatre edifice, there is a traditional precedent, since the earliest Greek theatre illustrated the center and was connected to the labyrinth and the symbolic dance; the circular theatre had an altar or the statue of a god

(which is a parody of the cathedral), by the Shopping Mall, and even by the "red light district," like in Amsterdam.¹

in the middle (see C. N. Deeds, *The Labyrinth*, in *The Labyrinth*, pp. 32-4; we may note that this collection of essays, written from the "historian of religions'" profane viewpoint, is worthy only for documentation, since the authors' perspective is an upside-down one, imbued with all the modern errors).

¹ For all this see Sedlmayr. Modern art, Sedlmayr said, brings forward the infrahuman, chaos and hell; and "surrealism" should be called in fact "sous-realism" (p. 143). And Sedlmayr rightly concluded: "that progress may quite well be a progress towards its end" (p. 241). Before ending this chapter and without elaborating too much, we would like to mention a "Darwinist" author, in fashion today, Richard Dawkins (for whom Darwin is a kind of god), to illustrate how the theory of "evolution" is desperately kept in place, despite its ridiculous premises, to enforce the idea of "progress." The main problem with "Darwinism" is its ignorance with regard to other domains than the corporeal one; its limitation to the material level guarantees its total failure. Dawkins scorns the "creationists," but they are no challenge, since both the "Darwinists" and "creationists" have a similar mentality; the author, of course, cannot even think about trying to confront the metaphysical immutable laws, a domain that is inaccessible to him. In his most recent book (*The Greatest Show on Earth*, Free Press, 2009), which is intended to be an "evidence for evolution," the author with a tedious persistence brings as proof the "natural selection," which, certainly, has nothing to do with the so called "evolution" (as the various races emerged from the primordial race, and, despite their differences, are all "human," so the "mutations" did not change a species or family into another one; also, there is no "natural selection" to explain why man evolved in becoming a defenceless, weak and unprotected being); Dawkins said: "the experiments ... show the power of natural selection to wreak evolutionary change" (p. 131), but the adjective "evolutionary" is gratuitous. The modifications are not part of a "natural selection"; they are a natural result of the metaphysical truth that the possibilities of manifestation are innumerable. There is today a change in tactics, since the promoters of evolution realized how ridiculous it was to pretend that man descended from monkeys; therefore, Dawkins used the idea of the "common ancestor," which is applied in a "regression to infinity," but which, in the end, meets the "creationists," since the first and primordial "common ancestor" cannot be anything else but a demiurge (even when the organic matter comes from the inorganic one). Also, Dawkins tried to change the meaning of "evolution" saying that the human being is not the "evolution's last word" (yet, in fact, nothing is new here, since in the past the evolutionists infected the young generations with ideas like the one about how man will have a huge brain and head and a tiny body, or how evolved nations will be without governments as rulers). However, when the author tried to explain why some animals left the

water for the land and then left the land for the water, the book lost any sense of reality, and plunged into a world of fantasy; the same fantasy characterizes the hypothesis that regards Africa as the origin of humankind (in the past, there were other places, like Pamir, considered to be this origin). The author's conclusion is: "there is no overall plan of development, no blueprint, no architect's plan, no architect" (p. 247), that is, there is only chaos, a conclusion based on Dawkins' ignorance and his knowledge limited to the creationist arguments. The Great Architect of the Universe is not an old man with long white beard sticking his nose in the world's problems, but this is too difficult to understand.

V

THE VISION OF THE CENTER

The Christian temple has two models: Heavenly Jerusalem, which is the center of our world or cycle, and Solomon's Temple,[1] which is a secondary center. Bede begins his *On the Temple* with this statement: "The house of God which king Solomon built in Jerusalem was made as a figure of the holy universal Church which, from the first of the elect to the last to be born at the end of the world, is daily being built through the grace of the king of peace, namely, its redeemer." As we already saw, for Gregory the Great Ezekiel's Temple and the city of Jerusalem represented equally Heavenly Jerusalem and the Church. Burckhardt specified that "the liturgy for the consecration of a church, the form of which can be traced back to the early Middle Ages, expressly compares the sacred edifice to the Heavenly Jerusalem"; in fact, the ecclesiastical architecture has taken its symbolism from both the Earthly Paradise and Heavenly Jerusalem.[2]

The question is: which was the exact temple taken as model and called "Solomon's Temple"?

---

[1] Abelard refers to Solomon's Temple as God's royal palace (His real temple is Heavenly Jerusalem). We may note that Solomon's royal palace was adjacent to the Temple, both marking the center.

[2] *Chartres*, pp. 24, 27. "On entering a newly built church, the hymn *Urbs Jerusalem beata* was sung: Blessed city of Jerusalem, known as vision of peace,/ Built in Heaven of living stones" (*Chartres*, p. 58).

There is no doubt that, as René Guénon underlined, the Temple of Jerusalem was a secondary center.[1] The institution of a spiritual center, Guénon said, is directly connected to the Divinity's "real presence," *Shekinah*, and the Scripture mentioned it mainly in relation to the construction of Solomon and Zerubbabel's temples.[2] The Temple (and especially the Holy of Holies), being a spiritual center, was the image of the Center of the World, which the Judaic Kabbalah described as the "Holy Palace" or the "inner Palace."[3] This Temple as a secondary center and above all as the image of the Center of the World was considered the model for the Christian church, and therefore the "holy place," regarded as a projection of the Center, was much more important than the physical temple itself, which explains why the church's prototype was at the same time Heavenly Jerusalem and Solomon's Temple, both "ideal" and "celestial" centers.[4]

Even though Moses was shown "heavenly things," Origen stressed, the Tabernacle and later the Temple were merely

---

[1] "Sion could be nothing else than one of these secondary centers" (*Le Roi*, p. 57). "Thus saith the Lord God; This is Jerusalem: I have set it in the midst of the nations and countries that are round about her" (*Ezekiel* 5:5). "And in the center was the temple itself, beautiful beyond all possible description, as one may conjecture from what is now seen around on the outside; for what is innermost is invisible to every human creature except the high priest alone, and he is enjoined only to enter that holy place once in each year" (Philo, *De Specialibus Legibus*, I, 72).

[2] *Le Roi*, pp. 22-23, *La Grande Triade*, p. 138, *Formes traditionnelles et cycles cosmiques*, p. 95, *Écrits*, p. 109; see also the second chapter of the present work.

[3] René Guénon compared the "Holy Palace" to *Brahmapura*, the "divine city" (*Écrits*, p. 106).

[4] We should recall the "invisible temple" of the Rose-Cross. Burckhardt said: "the Temple of Solomon is indeed the 'model' of the earthly body of the Lord, but the latter is, in a much truer sense than the temple of stone, the dwelling place of God" (*Chartres*, p. 22). René Guénon wrote that the symbolism of the medieval builders is based on the Judaic-Christian tradition and is especially attached to Solomon's Temple as the "prototype" (*Symboles fondamentaux*, p. 288). And Bede clearly stated that Christ is the "true Solomon" (Bede 20).

copies of the reality[1]; the visible sanctuary was the present Church on earth, while the invisible sanctuary (the Holy of Holies) is a figure of heaven itself.[2] As St. Augustine said, the divine promise made to King David about the building of the future Temple referred in fact to the temple made not of wood and stone, but of human beings.[3] There is a constant concern to emphasize the real archetype, the real goal, which is the absolute Center. As Bede avowed, the churches built down here and identified with the Temple of Jerusalem are mere copies of the New Jerusalem, since the true Church is the spiritual one.[4] Therefore, the real Solomon's Temple was identified with Ezekiel's vision, which was in a way comparable to Heavenly Jerusalem.[5]

Luc Benoist specified: "the church reminds us, through its forms and proportions, of Solomon's Temple, which Ezekiel saw in his dream, and, at the same time, the Heavenly Jerusalem."[6] If, at the beginning, a sanctuary had a cubic base

---

[1] As Peter of Celle has said: "let us visualize the historical Tabernacle that Moses built, as a means of guiding our attention from things visible to things invisible. Moses himself meant to direct the minds of the initiated, by means of the work he was building, to the spiritual vision that lay behind it. He, whose true tabernacle is in heaven, has nevertheless commanded the building of His sanctuary here on earth, so that the eye, illuminated by faith and reason, might perceive, 'as in a mirror and an enigma, the glory that lays beyond. Hence where should we treat of the tabernacle, if not in the tabernacle itself?'" (Simson 193). For Bede, the Temple and not the Tabernacle is a better image of the universal Church, since Solomon's Temple was built with foreign workers (Bede XXXIV, XXXV, XXXVIII, 6, 7, 8). Durantis said: "From both of these, namely, from the Tabernacle and the Temple, doth our material church take its form" (p. 20).
[2] Bede XXIV.
[3] Bede XXVI.
[4] Bede LI.
[5] Maximus the Confessor defined the "symbolic vision" as the capability "to apprehend within the objects of sense perception the invisibility reality of the intelligible that lays beyond them" (Simson XVII). In the time of the Christian traditional society, "the medieval mind was preoccupied with the symbolic nature of the world of appearances. Everywhere the visible seemed to reflect the invisible" (Simson XVIII).
[6] Luc Benoist, *Art du Monde*, Gallimard, 1941, p. 47.

crowned with a hemisphere to symbolize the Center,[1] later on the Romanesque and Byzantine churches became more explicit by using paintings and mosaics to suggest the "divine city" and make the believers forget the material temple and concentrate on the spiritual one. For the Masons who built the Gothic cathedrals the church was also an image of heaven, but now they tried to transform the material and earthly church itself into the Heavenly Jerusalem, replacing the stones with light[2]; as Simson said, "the Gothic may be described as transparent diaphanous architecture."[3]

As Simson so well has pointed out, the Heavenly Jerusalem was not the only model for the Christian temple; Solomon's Temple and the Temple of Ezekiel were also symbols of the "divine city," of the Center, and inspired the medieval Masons, and Solomon's Temple was viewed as an image of the Heavenly City.[4] At Schwarzrheindorf, Simson said, the Romanesque two-storied church was a combination of the two sources: "while

---

[1] It is interesting to notice how Hagia Sophia influenced the building of San Vitale, Ravenna, which, in its turn, influenced the architecture of the Palatine Chapel, Aachen (Charlemagne's "spiritual center").

[2] See Simson 8-11. "All important elements of Gothic architecture are almost literal representations of features of the Heavenly Jerusalem as described by St. John."

[3] Simson 4-5. "The Romanesque church building is earth in its lower reaches, Heaven in its height. Around the space in a Gothic church, Heaven itself descends like a mantle of crystalline light." The doctrine of Dionysius the Areopagite about the Divine Light was brilliantly applied to the construction of the cathedral of Chartres (Burckhardt, *Chartres*, pp. 35, 41-2). See Simson 50, 52, 103-106, 120-121 about the essential importance of light in Gothic style, the equivalence "between Dionysian light metaphysics and Gothic luminosity," and Suger's use of stained-glass windows as a visual application of Dionysian lore; St. Bernard saw union with God as "immersion in the infinite ocean of eternal light and luminous eternity" (Simson 123).

[4] Suger admitted that his Gothic abbey of St.-Denis had two sources: Hagia Sophia at Constantinople and Solomon's Temple, and he knew "that both sanctuaries were built with the same purpose in mind, and that both have the same author – God himself, since the Temple built by Solomon under divine guidance is also the model for the sanctuary" (Simson 95-6). When Hagia Sophia was completed, the emperor Justinian, it is said, exclaimed: "Solomon, I have surpassed you."

the murals of the upper church depict mainly scenes from the Book of Revelation, those of the lower church were inspired by the Vision of Ezekiel. The pictorial cycle in the lower church, one of the most impressive of the Middle Ages, seeks to make us see the entire sanctuary as the setting for Ezekiel's eschatological image."[1] The difference between the Romanesque and Gothic temple in representing the vision of heaven was that in the latter case the Geometry develops at its culmination,[2] which means the Operative Masonic Art reaches its peak.[3]

The description of the Heavenly Jerusalem is well known: "I John saw the holy city, new Jerusalem, coming down from God out of heaven. ... And he said unto me, Write: for these words are true and faithful. ... And he carried me away in the spirit to a great and high mountain, and shewed me that great city, the holy Jerusalem, descending out of heaven from God, having the glory of God (*Shekinah*): and her light was like unto a stone most precious, even like a jasper stone, clear as crystal[4]; and had a wall great and high, and had twelve gates, and at the gates twelve angels, and names written thereon, which are the names of the twelve tribes of the children of Israel. ... And he that talked with me had a golden reed to measure the city, and the gates thereof, and the wall thereof.[5] ... And the foundations of the wall of the city were garnished with all manner of precious stones. ... And I saw no temple therein: for the Lord God Almighty and the Lamb are the temple of it. And the city had no need of the sun, neither of the moon, to shine in it: for the glory of God did lighten it, and the Lamb is the light thereof.[6] ... And he shewed me a pure river of water of life, clear as crystal, proceeding out of the throne of God and of the Lamb.[7] In the

---

[1] Simson 9.
[2] Simson 13-14.
[3] We should note that the Art has a "non-human" origin (Benoist 30).
[4] We notice the symbolism of the stone.
[5] *Revelation* 21:2-12.
[6] *Revelation* 21:19, 21:22-23.
[7] The throne as center appears in *Isaiah*: "I saw also the Lord sitting upon a throne, high and lifted up, and his train filled the temple" (6:1); Isaiah's vision

midst of the street of it, and on either side of the river, was there the tree of life, which bare twelve manners of fruits, and yielded her fruit every month: and the leaves of the tree were for the healing of the nations."[1]

St. John described the same Center as the one revealed by Ezekiel's Vision[2]: "In the vision of God he brought me into the land of Israel, and set me upon a very high mountain,[3] by which was as the frame of a city on the south. And he brought me thither, and, behold, there was a man, whose appearance was like the appearance of brass, with a line of flax in his hand, and a measuring reed[4]; and he stood in the gate."[5] Ezekiel's vision continues with a very thorough and almost tedious series of measurements,[6] which are meant to present God as the Great

---

is that of the Heavenly Temple. "Thus saith the Lord, The heaven is my throne, and the earth is my footstool: where is the house that ye build unto me?" (*Isaiah* 66:1). As the Principle is the Center, so *Er-Rûh* is the Throne (Guénon, *Aperçus sur l'ésotérisme islamique*, p. 59).

[1] *Revelation* 22:1-2.

[2] We should quote here Ibn 'Arabî's sayings: "The vision of God on the Day of the Visit is according to men's beliefs in this world. Thus the person who believes concerning his Lord what was given to him by intellectual reflection (*nazar*), and by immediate 'unveiling' (*kashf*), and by imitating (*taqlîd*) his Messenger (*rasûl*) sees his Lord in the form of the aspect of each belief he held concerning Him" (*The Meccan Revelations*, I, p. 118).

[3] St. John said the same thing: "And he carried me away in the spirit to a great and high mountain" (*Revelation* 21:10).

[4] He is the divine Mason, yet we notice the brass, which alludes to a better Age. St. John said: "And he that talked with me had a golden reed to measure the city (*Revelation* 21:15). "I lifted up mine eyes again, and looked, and behold a man with a measuring line in his hand. Then said I, Whither goest thou? And he said unto me, To measure Jerusalem, to see what is the breadth thereof, and what is the length thereof" (*Zechariah* 2:1-2).

[5] *Ezekiel* 40:2-3.

[6] *Ezekiel* 40:5-15, 40:19-37, 40:47-49, 41:1-5, 41:8-15, 42:15-20, 43:13-17, 45:1-3. See also *Revelation* 21:16-18, where the art of measuring is described. "No medieval reader could have failed to notice with what emphasis every Biblical description of a sacred edifice, particularly those of Solomon's Temple, of the Heavenly Jerusalem, and of the vision of Ezekiel, dwells on the measurements of these buildings. To these measurements Abelard gives a truly Platonic significance. Solomon's Temple, he remarks, was pervaded by the divine harmony as were the celestial spheres" (Simson 37).

Architect of the Universe, to stress the fundamental importance of the proportions and geometry, to underline how real the vision is,[1] to indicate that the Great Architect has planned every detail and there is nothing haphazard about the construction of this sanctuary, and to remind us of the symbolism of measuring.[2] "And he brought me to the *Ulam* (porch) of the house, and measured each post of the porch, five cubits on this side, and five cubits on that side: and the breadth of the gate was three cubits on this side, and three cubits on that side.[3] ... Afterward he brought me into the *Hekal* (temple), and measured the posts, six cubits broad on the one side, and six cubits broad on the other side, which was the breadth of the tabernacle.[4] ... Then went he inward (the *Debir*), and measured the post of the door, two cubits; and the door, six cubits; and the breadth of the door, seven cubits. So he measured the length thereof, twenty cubits; and the breadth, twenty cubits, before the temple: and he said unto me, This is the most holy place.[5] ... Afterward he brought me to the gate, even the gate that looketh toward the east: And, behold, the glory of the God

---

[1] This reality is the divine one, the only real reality.
[2] See René Guénon and Ananda K. Coomaraswamy. The "measure," Guénon explained, symbolizes the actualization of the possibilities of manifestation, which describes the production of the world, of the universal manifestation as "order"; it is *Fiat Lux*. Coomaraswamy specified that Infinity is the "non-measurable" and what is "measured" represents what is "ordered," that is, the cosmos (Guénon, *Le règne*, pp. 42-43). St. Augustine's interpretation of the famous passage "thou hast ordered all things in measure and number and weight" (*Wisdom of Solomon* 11:20) became the keyword of the medieval world's traditional view with respect to architecture, music, or poetry (Simson 25), and Guénon reminded us of this very Biblical passage, which is an expression of the universal "order" (*Le règne*, p. 45). Also, René Guénon underlined that "measure" is directly connected to a symbolic and initiatory "geometry," and on this is based all the conceptions that assimilate the divine activity of producing and ordering the worlds with "geometry," and, consequently, with "architecture"; all these are related to Plato's sayings, "God forever geometrizes" (*Le règne*, pp. 45-46). The correct measure is directly related to justice, and justice is a characteristic of the center.
[3] *Ezekiel* 40:48.
[4] *Ezekiel* 41:1.
[5] *Ezekiel* 41:3-4.

of Israel (*Shekinah*) came from the way of the east: and his voice was like a noise of many waters: and the earth shined with his glory.[1] ... And the glory of the Lord came into the house by the way of the gate whose prospect is toward the east. So the spirit took me up, and brought me into the inner court; and, behold, the glory of the Lord (*Shekinah*) filled the house.[2] ... Afterward he brought me again unto the door of the house; and, behold, waters issued out from under the threshold of the house eastward: for the forefront of the house stood toward the east, and the waters came down from under from the right side of the house, at the south side of the altar. ... And when the man that had the line in his hand went forth eastward, he measured a thousand cubits, and he brought me through the waters; the waters were to the ankles. Again he measured a thousand, and brought me through the waters; the waters were to the knees. Again he measured a thousand, and brought me through; the waters were to the loins. Afterward he measured a thousand; and it was a river that I could not pass over: for the waters were risen, waters to swim in, a river that could not be passed over. And he said unto me, Son of man, hast thou seen this? Then he brought me, and caused me to return to the brink of the river. Now when I had returned, behold, at the bank of the river were very many trees on the one side and on the other. Then said he unto me, These waters issue out toward the east country, and go down into the desert, and go into the sea: which being brought forth into the sea, the waters shall be healed."[3]

---

[1] *Ezekiel* 43:1-2.
[2] *Ezekiel* 43:4-5.
[3] *Ezekiel* 47:1-8. St. John said: "And he showed me a pure river of water of life, clear as crystal, proceeding out of the throne of God and of the Lamb. In the midst of the street of it, and on either side of the river, was there the tree of life" (*Revelation* 22:1-2). The "river of life" described by Ezekiel and St. John should be compared to the Hindu river Gangâ, which fell from heaven into Shiva's hair and from there she became an earthly holy river flowing into the ocean. In the Islamic and post-Vedic traditions the guardian of the "river of life" is the mysterious Al-Khadir (Khwâjâ Khizr) (see Coomaraswamy, *What is Civilisation?*, chapter *Khwâjâ Khadir and the Fountain of Life in the Tradition of*

The detailed measurements given by God, we must insist, constitute the "explication" of the divine vision on earth and symbolize how the divine inspiration operates in the world. Using a divine archetype to build a terrestrial temple, city or palace meant to apply effectively the art and science of the Great Architect of the Universe, to obey God's command regarding the "measurements," which implied that, following the heavenly proportions and mastering the art of geometry and the science of rhythm, the human architect could be certain that the edifice would be perfect. At the construction of the cathedral of Milan, the French architect Jean Mignot, who in 1398 pronounced the famous statement *ars sine scientia nihil*, warned that if the proportions are not obeyed the building would fall.[1] What Mignot tried to suggest was the enormous difference between the traditional and sacred art of Masonry

---

Persian and Mughal Art, previously published in Études Traditionnelles, 1938, no. 224-225); Moses encounters Al-Khadir on a rock marking the meeting of the two seas (*Qur'an* 18:60-82), a rock on which the Temple (the "Dome of the Rock") will be built, a rock that Hesychasm compares to the hardened heart of the man – a heart that must be split to liberate the waters made of tears (equivalent to the "river of life"); "Then after that your hearts became hard like rocks, or harder still: for verily, from rocks have rivers gushed; others, verily, have been cleft, and water hath issued from them" (*Qur'an* 2:72-74) (see Martin Lings, *Le symbolisme coranique de l'eau*, Études Traditionnelles, 1970, no. 421-422). "He is the One who has cut off the two seas, this one being sweet, fresh, while the other is salty, bitter. He has placed an isthmus (*barzakh*) in between them, and a barrier that cannot be passed" (*Qur'an* 25:53); as Titus Burckhardt stated, *barzakh* is, in Sufism, an equivalent to the pole (*qutb*) (*Symboles*, p. 86) and René Guénon specified that the Heart of the World (the luminous sphere which is *Rûh mohammediyah*) is the sole true *barzakh* (*Symboles fondamentaux*, p. 228). If, on the one hand, the "river of life" springs from the central "temple," on the other hand, the "temple" is the isthmus between the two seas, as the center is the contact point between Heaven and Earth (Guénon, *Écrits*, p. 112) (the Sumerian temples at Nippur and Larsa where called *Dur-an-ki*, "bond of Heaven and Earth"); the Judaic tradition considered Mount Sion the center and called it the "Heart of the World" (Guénon, *Le Roi*, p. 56). Finally, we should add that Jerusalem was the place where someone praying for rain should obtain it, since, as the Islamic tradition affirms, "there is no sweet water of which the source does not originate under the Holy Rock of Jerusalem" (Wensinck, *The Navel*, p. 33).

[1] Burckhardt, *Chartres*, p. 102, Simson 19.

and the profane one: the construction of an edifice in a traditional society was conducted starting from the divine principles, complying with the metaphysical doctrine, with the principles of rhythm, proportions, and harmony,[1] striving to realize the perfect equilibrium of the edifice (*coincidentia oppositorum*), to build "according to *true measure*,"[2] and as secondary consequences the building satisfied all the laws and rules of Physics and Mechanics (as understood in a modern sense); on the contrary, the modern and profane builder ignores the divine vision and uses an analytical and experimental approach, with all kinds of profane calculations (strength of materials, etc.), which make the edifice a vain, ugly and insecure result.[3] As the French architect of the 16th Century Philippe Delorme said, God is the great and admirable architect, who by creating the cosmos according to measure and number and weight has also given to the human architect proportions so perfect that without such divine help he would never have been able to discover them.[4]

As René Guénon said, in the Middle Ages, certain initiatory organizations identified the seven "liberal arts" with various degrees of initiation, which is not surprising since the traditional métiers were also supports for spiritual realization, and, in fact,

---

[1] A traditional *chef d'œuvre* is an assembly of symbols supported by rhythms (Benoist 29). Boethius, St. Augustine's disciple, established the "geometrical harmony" of the cube (Simson 33).

[2] Simson 16.

[3] A traditional work of art, like a cathedral or a sacred painting, like the *Iliad* or the *Divine Comedy*, is a *chef d'oeuvre* not because it produces a supreme delectation of our senses, because it is beautiful, but because, being traditional and following divine inspiration, it is perfect, and it implicitly contains all the secondary characteristics. As Coomaraswamy said, art is intellectual and not esthetic (Coomaraswamy, *Christian and Oriental Philosophy of Art*, p. 16).

[4] Simson 227-228. Delorme recalled the "sacred proportions" that God prescribed to Noah (for the Ark), to Moses (for the Tabernacle), to Solomon (for the Temple), and to Ezekiel (for the Celestial Temple). Saint Irenaeus, in his *Adversus haereses*, wrote: "God makes all things with measure and order, and nothing lacks measure, since it does not lack number" (4.4).

The vision of the Center 127

there was no difference between art and métier.[1] Traditional art had nothing to do with modern art, which is viewed as a provider of pleasure and emotions[2]; any traditional art was an application of the science of rhythm,[3] related to the science of number (which is not the profane arithmetic, of course, but the science used by Kabbalah and Pythagorism[4]), and therefore traditional architecture was closely connected to music, both using the same divine proportions and rhythms.[5] In this regard, St. Augustine considered music and architecture to be sisters, both based on perfect ratios, and all artistic creations observe the laws of numbers.[6] The influence of Pythagoras and Plato is to be noticed in the two main "schools" related to the Gothic

---

[1] See Guénon, *Mélanges*, pp. 102-4. Expressions like "Sacerdotal Art" and "Royal Art" illustrate Mignot sayings and show what the traditional art really was. As Thierry of Chartres said, "philosophy [understood as "love for wisdom"] has two main instruments, namely intellect (*intellectus*) and its expression. Intellect is illuminated by the *quadrivium* (Arithmetic, Music, Geometry and Astronomy). Its expression is the concern of the *trivium* (Grammar, Dialectic and Rhetoric)" (Burckhardt, *Chartres*, p. 70). As the natural phenomena and the historical events have all a symbolic value, Guénon insisted, so "any art and science could, using a convenient transposition, acquire an esoteric value"; therefore, the "liberal arts" also could play an initiatory role in the Middle Ages, as the symbols related to construction and architecture play a similar role in the present Masonry. The "liberal arts" correspond to the first seven heavenly circles of Dante and to the seven steps of the *Ladder of the Kadosch* (30th degree of the Scottish Masonry) (Guénon, *L'ésotérisme de Dante*, pp. 13-14, Guénon, *Aperçus sur l'ésotérisme chrétien*, pp. 59, 78).
[2] Benoist 17-18; Guénon, *Mélanges*, pp. 104-105.
[3] The science of rhythm was one of the most used means in the process of spiritual realization (René Guénon, *La métaphysique orientale*, Éditions Traditionnelles, 1993, p. 23); rhythms help the individual to communicate with the superior states of the being (Guénon, *Mélanges*, pp. 106-108; Benoist 29).
[4] Guénon, *L'ésotérisme de Dante*, p. 15.
[5] Guénon, *Mélanges*, pp. 102-108. Benoist 48, 54, Burckhardt, *Chartres*, p. 94. For Suger, the universe was "perfect music" (Burckhardt, *Chartres*, p. 42).
[6] Simson 21-24. Simson considers that St. Augustine's views regarding the divine proportions and his interpretation of the passage from the *Wisdom of Solomon* ("thou hast ordered all things in measure and number and weight") shaped the Middle Ages. However, what Simson overlooks is that, besides exotericism, the medieval esoteric knowledge was much more essential and influent.

period, that of the great St. Bernard[1] and that of Chartres, where Geometry and the science of numbers constituted the bridge between God and the world, where Theology became Geometry,[2] and God the Divine Architect.[3]

The science of rhythm was applied not only to "measurements" in connection to space, but also to those in connection to time. Ezekiel (like other prophets) is very thorough in presenting not only the spatial elements but also the temporal ones; his Book starts like this: "Now it came to pass in the thirtieth year, in the fourth month, in the fifth day of the month, as I was among the captives by the river of Chebar, that the heavens were opened, and I saw visions of God. In the fifth day of the month, which was the fifth year of king Jehoiachin's captivity,"[4] and the Vision of the Temple begins with: "In the five and twentieth year of our captivity, in the beginning of the year, in the tenth day of the month, in the fourteenth year after that the city was smitten, in the selfsame day the hand of the Lord was upon me, and brought me thither."[5] Dante commenced his initiatory poem, *La Divina Commedia*, in a similar way: "In the midway of this our mortal life,/ I found me in a gloomy wood, astray/ Gone from the path direct." René Guénon, commenting on Dante's work, said: Dante places his vision precisely in the middle of the "grand year," which is the middle of time[6]; but he also starts his initiatory voyage placed in a spatial point that is indeed the center of terrestrial world, this central point corresponding geographically to Jerusalem, which for Dante represented the "spiritual pole."[7] And Guénon added: "this point is both with respect to time and space [the Center of the World], that is,

---

[1] René Guénon considered St. Bernard an initiate. It is important to note St. Bernard's profound interest in music (Simson 41-42).
[2] See Simson 33-35 about Geometry (identical to Masonry) as art and science.
[3] Simson 25-29, 31.
[4] *Ezekiel* 1:1.
[5] *Ezekiel* 40:1.
[6] *L'ésotérisme de Dante*, pp. 60, 63.
[7] *L'ésotérisme de Dante*, p. 64.

with respect to the two conditions that essentially characterize existence in this world."[1]

Both of Ezekiel's visions are introduced by a meticulous description of the temporal coordinates, this similarity suggesting a more profound one: the vision of the divine chariot[2] and the vision of the Temple are in reality two illustrations of the same unique supreme Center. This means that, despite all the different opinions regarding the second vision, the prophet was referring primarily not to a material temple, but to a celestial one, a truth well understood by the Christian tradition, and also by the Operative Masons and the Order of the Temple.[3]

The Vision of Ezekiel was interpreted as alluding to the Millennial Temple, the Messianic kingdom, the future Christian Church, the temple that would be built after the Babylonian captivity, or to a new holy site, different from the Temple Mount site, with a "memorial" temple. There is a persistent opinion that Ezekiel was referring to a material temple that most probably could be the "millennial" temple, but which is definitely different from the Heavenly Jerusalem described by St. John (since the New Jerusalem does not include a temple: "And I saw no temple therein: for the Lord God Almighty and the Lamb are the temple of it"[4]), and Zechariah described precisely this material "millennial" temple.[5]

What Zechariah, like Ezekiel and St. John, portrayed was the Center: "Thus saith the Lord; I am returned unto Sion, and will

---

[1] *L'ésotérisme de Dante*, p. 66. Dante's initiatory voyage takes place during the Easter week (*L'ésotérisme de Dante*, p. 39); Ezekiel's vision occurs just before Passover, since: "In the first month, on the fourteenth day of the month, you shall have the Passover, a feast of seven days" (*Ezekiel* 45:21) and the vision appears "in the beginning of the year, in the tenth day of the month" (Passover is celebrated at the beginning of the religious year).
[2] See the first chapter of the present work.
[3] Therefore, all the New Age literature that tries to promote a material temple hiding documents and treasures, in connection to the Knights Templar, is just a subversive attempt to confuse peoples' mind.
[4] *Revelation* 21:22.
[5] Some authors even established the physical location of the "millennial" temple.

dwell in the midst of Jerusalem: and Jerusalem shall be called a city of truth; and the mountain of the Lord of hosts the holy mountain."[1] "Then shall the Lord go forth, and fight against those nations, as when he fought in the day of battle. And his feet shall stand in that day upon the mount of Olives, which is before Jerusalem on the east, and the mount of Olives shall cleave in the midst thereof toward the east and toward the west, and there shall be a very great valley; and half of the mountain shall remove toward the north, and half of it toward the south. ... And the Lord my God shall come, and all the saints with thee. And it shall come to pass in that day, that the light shall not be clear, nor dark: but it shall be one day which shall be known to the Lord, not day, nor night: but it shall come to pass, that at evening time it shall be light. And it shall be in that day, that living waters shall go out from Jerusalem[2]; half of them toward the former sea, and half of them toward the hinder sea: in summer and in winter shall it be. And the Lord shall be king over all the earth: in that day shall there be one Lord, and his name one."[3]

The mission to prophesy that the physical Temple of Jerusalem was a secondary center belonged to Haggai, a contemporary of Zechariah, born like him during the Babylonian captivity: "In the seventh month, in the one and twentieth day of the month, came the word of the Lord by the prophet Haggai, saying, Speak now to Zerubbabel the son of Shealtiel, governor of Judah, and to Joshua the son of Josedech, the high priest, and to the residue of the people, saying, Who is left among you that saw this house in her first glory? And how do ye see it now? Is it not in your eyes in comparison of it as nothing? Yet now be strong, O Zerubbabel, saith the Lord; and be strong, O Joshua, son of Josedech, the high priest; and be strong, all ye people of the land, saith the Lord, and work: for I am with you, and I will fill this house with glory, saith the Lord of hosts. The glory of this latter house [*Shekinah*] shall be greater

---

[1] *Zechariah* 8:3.
[2] Ezekiel and St. John mentioned the same "living waters."
[3] *Zechariah* 14:4-9.

than of the former, saith the Lord of hosts: and in this place will I give peace [the attribute of the center], saith the Lord of hosts."[1] Zechariah also referred to Zerubbabel: "Moreover the word of the Lord came unto me, saying, The hands of Zerubbabel have laid the foundation of this house; his hands shall also finish it; and thou shalt know that the Lord of hosts hath sent me unto you. For who hath despised the day of small things? For they shall rejoice, and shall see the plummet in the hand of Zerubbabel."[2]

Free-Masonry paid close attention to the construction of the second temple, to Zerubbabel and the other main characters, but what it had primarily in view was the symbolic meaning of the journey toward the center and of the foundation of a secondary center, as image of the supreme Center, which constituted for Zechariah the main theme of his apocalyptic prophecy: "Therefore thus saith the Lord; I am returned to Jerusalem with mercies: my house shall be built in it, saith the Lord of hosts, and a line shall be stretched forth upon Jerusalem. Cry yet, saying, Thus saith the Lord of hosts; My cities through prosperity shall yet be spread abroad; and the Lord shall yet comfort Sion, and shall yet choose Jerusalem."[3]

As we mentioned in the previous chapter, the two ends of the present *Manvantara* are, from a Christian perspective, Earthly Paradise (the Center with respect to the beginning of this cycle) and the Heavenly Jerusalem (the Center regarding the end of this world), where the evolvement of our cycle could be perceived as the process of transforming the circle into a square.[4] Earthly Paradise and Heavenly Jerusalem are two states of the same supreme Center, where after the termination of the cycle, the latter takes the place of the former and becomes Paradise for the new cycle.[5] Therefore, prophets like Ezekiel, Zechariah, or St. John related in a natural way the Center to the

---

[1] *Haggai* 2.
[2] *Zechariah* 4:8-10.
[3] *Zechariah* 1:16-17.
[4] Guénon, *Le règne*, pp. 191, 218.
[5] Guénon, *Le règne*, p. 192.

end of our world (since prophesizing regards the future not the past), but they also paid particular attention to the four "corner–stones," which have to be reabsorbed into the Center and which would play an important role to complete the exhaustion of the cycle, since "each of the four 'corner–stones' reflects the dominant principle [the Center]"[1] and "the quaternary was everywhere and always considered the number of the universal manifestation."[2]

The Masonic *dictum* "to reassemble what was scattered"[3] is not only an illustration of the initiatory process, but part of the "cosmic process as a whole: Purusha 'having been one, becomes many, and having been many, becomes again one,'"[4] René Guénon echoing here Coomaraswamy's words: "In sacrificing himself in the beginning, the Solar Hero, having been single, makes himself – or is made to be – many" and "having been many, he must again become one. ... The second phase of the Sacrifice ... consists in the putting together again of what

---

[1] Guénon, *Symboles fondamentaux*, p. 280. In addition, Guénon under-scored that *shethiyah*, the "foundation–stone" placed in the center of the square (representing the base of the edifice), synthesizes in itself the partial aspects represented by the four "corner–stones," while the "foundation–stone" and the "angle–stone" could describe different situations of the same center, similar to the various positions of *luz*, the "kernel of immortality," and comparable with the sundry Hindu *chakras* (*ibid.* p. 296). Guénon also noted the "method of the five points" used in Operative Masonry, where the location of an edifice was determined, before starting the construction, by fixing the four corners and the center of the base, the five points being the *landmarks* (*ibid.* p. 295).
[2] Guénon, *Symboles fondamentaux*, p. 125.
[3] Guénon, *Symboles fondamentaux*, p. 301; *Jeremiah* 31:10 ("Hear the word of the Lord, O ye nations, and declare it in the isles afar off, and say, He that scattered Israel will gather him, and keep him, as a shepherd doth his flock").
[4] Guénon, *Symboles fondamentaux*, p. 303. Maximus the Confessor, inspired by Dionysius, said: "It is he [Christ] who encloses in himself all beings by the unique, simple, and infinitely wise power of his goodness. As the center of straight lines that radiate from him he does not allow by his unique, simple, and single cause and power that the principles of beings become disjointed at the periphery but rather he circumscribes their extension in a circle and brings back to himself the distinctive elements of beings which he himself brought into existence" (p. 187).

had been dismembered. ... This unification and 'coming into one's own' is at once a death, a rebirth, an assimilation, and a marriage."[1] Guénon related this unification and reassembly to the myth of Osiris, but he also stressed the similarity with the Masonic legend regarding Hiram's sacrifice,[2] since, as Coomaraswamy stated, the Sacrifice has two facets: the first is the division of Prajâpati, who "pours out his offspring, makes himself many and enters into us," and doing so is "emptied out"; the second is when "the sacrificer draws in these breaths with Om, and sacrifices them in the Fire without evil," that is, "the sacrificer as it were emanates offspring and it thereupon emptied out as it were."[3] "The initiate," Coomaraswamy added, "assembles and would collect the Sacrifice; his self, as it were, is emptied out,"[4] which is comparable to death (or the state of profound sleep, or *samâdhi*) "when the breaths unite with it" and "the departing Breath ... extracts or impresses the breaths, as a horse might tear out the pegs."[5]

To die means "to breathe your last breath" or "to expire," which from a macrocosmic viewpoint signifies to end the world and, since the breath is comparable to the wind, it means to expire the winds and be emptied out. As Guénon and Coomaraswamy pointed out, a spiritual realization represents a death and a rebirth, when the initiate passes through the "eye of the dome," but for the ignorant this "solar door" is seen as the jaws of Death, since "his self, as it were, is emptied out." Therefore, "to reassemble what was scattered," corresponds to the initiatory [Self, Breath, Wind, Spirit] point of view, while from the cowan's [ego, breath, self's] perspective we have the reverse process, that is, to be "scattered by the four cardinal winds of heaven,"[6] and, transferred to the macrocosmic

---

[1] *Metaphysics*, pp. 136-7.
[2] *Symboles fondamentaux*, p. 302.
[3] Coomaraswamy, *Metaphysics*, p. 120.
[4] The expression "to collect himself" illustrates exactly the passing from multiplicity to unity.
[5] *Metaphysics*, p. 121. The Breath is Prajâpati, the Lord, and the breaths are his children, powers of the soul, the subjects of the king (pp. 122, 127).
[6] *Morgan's Expose Free Masonry*, p. 40, n.d. n.p.

standpoint, we could say that the end of the world occurs when the winds devastate the world and, at the same time, are reabsorbed and pacified into the Center, where, as Lie Zi stated, describing the Center, the wind does not blow.[1] Similarly, the Islamic tradition mentions more than once the apocalyptical function of the wind when describing the punishment of the tribe of Ad.[2]

---

[1] *The Book of Lieh-tzû*, Columbia University Press, 1990, p. 102. In Syriac literature it is said: "There [in Jerusalem] the four parts of the world have been united one with the other. When God made the earth, his power went before, and the earth followed from four sides swift as the winds; and there [in Jerusalem] his power stood and rested" (Wensinck, *The Navel*, p. 17). In the Romanian traditional vestiges, the center is similarly described as the place where the wind does not blow: the Tradition symbolized by a maiden is hidden in "a profound valley, in solitude, under stony slabs, where the wind doesn't blow, and nobody can see her," which alludes to a subterranean center; yet the same center is described as being located "on the highest mountain/ Where the wind doesn't blow/ And the dogs are not barking/ And the roosters are not singing." These portrayals are perfectly similar to the Masonic formula, regarding the Lodge's location: "on the highest mountain and in the deepest valley, which is the Valley of Jehoshaphat, and in any secret and silent location where nobody can hear the dog barking and the rooster singing" (Denys Roman, *Réflexions d'un chrétien sur la Franc-Maçonnerie*, Éd. Traditionnelles, 1995, p. 125). See our work *Agarttha, the Invisible Center*, Rose-Cross Books, 2002, pp. 20-21.

[2] Among the famous tribes of the ancient Arabs, besides Ad, we should mention those of Thamûd, Tasm, Jadîs, and Amalek. Ad was the grandson of Aram, the son of Sem (Shem), the son of Noah. As Solomon continued his father's plan to build the Temple, so Shedâd, the son of Ad, continued his father's construction of a paradisiacal center (city, palace and the garden of Irem, which was considered the substitute of Paradise); yet the human factor became so powerful and arrogant that the tribe of Ad forgot about the real and unique God, about the real Center, and fell into idolatry, confusing Paradise with the garden of Irem (which was just an image). God sent the prophet Hûd to save them, but they did not accept him, and therefore God sent a hot and mortal wind that blew for a week and destroyed them all. It is said that the ancient Adites were of prodigious stature, which could allude to Atlantis. "And unto (the tribe of) Ad (We sent) their brother, Hûd (*Qur'an* 7:65); "And in (the tribe of) Ad (there is a portent) when we sent the fatal wind against them" (51:41). For the traditional commentaries see those compiled by George Sale in *The Koran*, Frederic Warne and Co., n.d. (no later than 1853).

The vision of the Center 135

In Greek mythology, the four winds, related to the cardinal points, were famous: Boreas – the north wind, Eurus – the east wind, Notus – the south (or southwest) wind, and Zephyrus – the west wind.[1] Boreas was sometimes presented as a dark stallion and Zephyrus as the father of Achilles' immortal horses, Xanthus and Balius. This equivalence of wind – horse mirrors the equivalence of breath – horse from the previously quoted Hindu text. Zechariah mentioned the same four winds as chariots drawn by horses: "And I turned, and lifted up mine eyes, and looked, and, behold, there came four chariots out from between two mountains; and the mountains were mountains of brass.[2] In the first chariot were red horses; and in the second chariot black horses; and in the third chariot white horses; and in the fourth chariot grisled and bay horses. Then I answered and said unto the angel that talked with me, What are these, my lord? And the angel answered and said unto me, These are *the four winds of the heavens*,[3] which go forth from standing before the Lord of all the earth. The black horses which are therein go forth into the north country; and the white go forth after them; and the grisled go forth toward the south country. And the bay went forth, and sought to go that they might walk to and fro through the earth: and he said, Get you hence, walk to and fro through the earth. So they walked to and fro through the earth."[4]

---

[1] Charbonneau-Lassay, analyzing the keystone of what was left of the Cathedral of Mâcon, mentioned that, in the medieval iconography, Christ was represented as a lion in the center, while the four corners were human masques symbolizing the four winds Aquilo (Boreas), Eurus, Zephyrus, and Auster (Notus) *(La clef de voûte du narthex de Saint-Vincent de Mâcon*, Études Traditionnelles, no. 357, 1960, p. 9; see some more details about the four winds in René Mutel, *A propos de la clef de voûte mâconnaise au lion bibliophore*, ET, no. 358, pp. 69 ff.). It is interesting that in the Grail stories Christ is the white hart and in the corners are four lions (*The Quest of the Holy Grail*, Penguin Books, 1969, p. 243).
[2] This is the "solar door"; the brass alludes to a better Age.
[3] Our italic. In the Ancient Egypt, as part of the "doctrine of the gesture," ritual dances, illustrating the four winds, were performed.
[4] *Zechariah* 6:1-7.

The four types of horses[1] are found again in St. John's apocalyptic *Revelation*: "And I saw when the Lamb opened one of the seals, and I heard, as it were the noise of thunder, one of the four beasts saying, Come and see. And I saw, and behold a white horse: and he that sat on him had a bow; and a crown was given unto him: and he went forth conquering, and to conquer. And when he had opened the second seal, I heard the second beast say, Come and see. And there went out another horse that was red: and power was given to him that sat thereon to take peace from the earth, and that they should kill one another: and there was given unto him a great sword. And when he had opened the third seal, I heard the third beast say, Come and see. And I beheld, and lo a black horse; and he that sat on him had a pair of balances in his hand. And I heard a voice in the midst of the four beasts say, A measure of wheat for a penny, and three measures of barley for a penny; and see thou hurt not the oil and the wine. And when he had opened the fourth seal, I heard the voice of the fourth beast say, Come and see. And I looked, and behold a pale horse: and his name that sat on him was Death, and Hell followed with him. And power was given unto them over the fourth part of the earth, to kill with sword, and with hunger, and with death, and with the beasts of the earth."[2]

It is hard to understand why exactly the rider of the white horse was identified with the Antichrist, but as an Arthurian legend said, Satan apes God and dresses in white, and therefore the first of the four apocalyptic horsemen is somehow the ape of Christ, who is the absolute rider of the white horse: "And I saw heaven opened, and behold a white horse; and he that sat upon him was called Faithful and True, and in righteousness he doth judge and make war. His eyes were as a flame of fire, and on his head were many crowns; and he had a name written, that no man knew, but he himself. And he was clothed with a vesture dipped in blood: and his name is called The Word of

---

[1] The four colours are related to the four colours (Sanskrit *varna*) of the Hindu casts (Guénon, *Études sur l'hindouisme*, p. 77) and of the four Ages.
[2] *Revelation* 6:1-8.

The vision of the Center 137

God.[1] And the armies which were in heaven followed him upon white horses, clothed in fine linen, white and clean. And out of his mouth goeth a sharp sword, that with it he should smite the nations: and he shall rule them with a rod of iron: and he treadeth the winepress of the fierceness and wrath of Almighty God. And he hath on his vesture and on his thigh a name written, King of Kings, and Lord of Lords."[2] This vision is similar to Zechariah's vision: "Upon the four and twentieth day of the eleventh month, which is the month Sebat, in the second year of Darius, came the word of the Lord unto Zechariah, the son of Berechiah, the son of Iddo the prophet, saying, I saw by night, and behold a man riding upon a red horse, and he stood among the myrtle trees that were in the bottom[3]; and behind

---

[1] We can find in the Grail stories the same sacred and initiatory symbolism of the name.

[2] *Revelation* 19:11-16. It is possible that *Kalki-avatâra*, riding a white horse and armed with a sword and bow, influenced the assimilation of the first horseman with Antichrist. However, Jean Reyor underlined that René Guénon was probably the first to mention that "*Kalki-avatâra*, 'the rider of the white horse' and the one who must come at the end of this cycle, is described in the *Purânas* in terms rigorously identical to those found in the *Revelation*, where they refer to the 'second coming' of Christ" (Jean Reyor, *Études et recherches traditionnelles*, Éditions Traditionnelles, 1991, p. 52, Guénon, *Symboles fondamentaux*, p. 168, Martin Lings, *Ancient Beliefs and Modern Superstitions*, Suhail Academy Lahore, 1999, p. 24). As André Préau affirmed, the white horse became identical to *Kalki-avatâra* and various authors noticed the resemblance with the apocalyptic white horse (*Le Kalki-Purâna*, Archè, 1982, p. 195).

[3] Zechariah's rider is similar to the Saviour on the white horse from St. John's description. The fact that Zechariah's horse has a red colour could allude to the malefic significance of the "red donkey," in which case the rider is subduing the malefic forces by riding them. Guénon gave the same interpretation with regard to Christ riding the donkey, an image present both in Zechariah and the Gospel: "Rejoice greatly, O daughter of Sion; shout, O daughter of Jerusalem: behold, thy King cometh unto thee: he is just, and having salvation; lowly, and riding upon a donkey, and upon a colt the foal of a donkey" (*Zechariah* 9:9); "All this was done, that it might be fulfilled which was spoken by the prophet, saying, Tell ye the daughter of Sion, behold, thy King cometh unto thee, meek, and sitting upon an ass, and a colt the foal of an ass. And the disciples went, and did as Jesus commanded them, and brought the ass, and the colt, and put on them their clothes, and they set him

him were there red horses, speckled, and white. ... These are they whom the Lord hath sent to walk to and fro through the earth."[1]

We could summarize all this in a synthetic picture having the Knight on the white (red) horse (the Breath) in the center and the other four horsemen (the four winds or breaths) at the four cardinal points or in the four corners. From a mundane viewpoint, the four apocalyptic horsemen represent malefic forces, terrible Death, even the Antichrist; they were compared with Daniel's four beasts: "In the first year of Belshazzar king of Babylon Daniel had a dream and visions of his head upon his bed: then he wrote the dream, and told the sum of the matters. Daniel spake and said, I saw in my vision by night, and, behold, *the four winds of the heaven*[2] strove upon the great sea. And four great beasts came up from the sea, diverse one from another. The first was like a lion, and had eagle's wings: I beheld till the wings thereof were plucked, and it was lifted up from the earth, and made stand upon the feet as a man, and a man's heart was given to it. And behold another beast, a second, like to a bear, and it raised up itself on one side, and it had three ribs in the mouth of it between the teeth of it: and they said thus unto it, Arise, devour much flesh. After this I beheld, and lo another, like a leopard, which had upon the back of it four wings of a fowl; the beast had also four heads; and dominion was given to it. After this I saw in the night visions, and behold a fourth beast, dreadful and terrible, and strong exceedingly; and it had great iron teeth: it devoured and broke in pieces, and stamped the residue with the feet of it: and it was diverse from all the beasts that were before it; and it had ten horns. I considered the horns, and, behold, there came up among them another little horn, before whom there were three of the first horns plucked

---

thereon" (*Matthew* 21:4-7). With respect to the ambivalence of the red colour's symbolism see the Grail stories.
[1] *Zechariah* 1:7-8, 1:10.
[2] Our italic.

up by the roots: and, behold, in this horn were eyes like the eyes of man, and a mouth speaking great things."[1]

Even though Daniel's four beasts have a terrifying appearance and obviously their function is to destroy and dissolve, they are ultimately the guardians of the four cardinal points, since at the beginning of the vision they are identified with "the four winds of the heaven [that] strove upon the great sea." René Guénon and Ananda K. Coomaraswamy extensively explained the symbolism of the deadly jaws and of the monstrous guardians; Daniel's four beasts belong to the same symbolic family and so do the four apocalyptic horsemen, who are also the guardians of the cardinal points or of the corners of the earth, where their function is not so much to destroy (a mundane viewpoint, when the jaws are the Death's jaws), but to transform (when the jaws are the "sun door"), as in Shiva's case. The four apocalyptic horsemen relate to the four cardinal points (and, in a way, to the four Ages of the cycle), since each knight is announced by a figure of the tetramorph: "And before the throne there was a sea of glass like unto crystal: and in the midst of the throne, and round about the throne, were four beasts full of eyes before and behind. And the first beast was like a lion, and the second beast like a calf, and the third beast had a face as a man, and the fourth beast was like a flying eagle."[2] Moreover, it is said: "And after these things I saw four angels standing on the four corners of the earth, holding *the four winds of the earth*,[3] that the wind should not blow on the earth,

---

[1] *Daniel* 7:1-8. "Blow ye the trumpet in Sion, and sound an alarm in my holy mountain: let all the inhabitants of the land tremble: for the day of the Lord cometh, for it is nigh at hand; A day of darkness and of gloominess, a day of clouds and of thick darkness, as the morning spread upon the mountains: a great people and a strong; there hath not been ever the like, neither shall be any more after it, even to the years of many generations. A fire devoureth before them; and behind them a flame burneth: the land is as the garden of Eden before them, and behind them a desolate wilderness; yea, and nothing shall escape them. The appearance of them is as the appearance of horses; and as horsemen, so shall they run" (*Joel* 2:1-4).
[2] *Revelation* 4:6-7.
[3] Our italic.

nor on the sea, nor on any tree."[1] The last quotation alludes to the Center, situated beyond the cycle and the world; it is the Center where the wind does not blow; it is the Center where, after they have dissolved the world, the four winds unite and are pacified.

As we saw in the first chapter, the tetramorph is present in Ezekiel's opening vision of God (Center) as universal vortex; as primordial luminous sphere with its length, width, and depth delineating the three-dimensional cross; as spatial *swastika* – symbol of the Pole, which has on a horizontal plane (as it is usually drawn) four branches associated with the four creatures. The Chaldeans had their own similar tetramorph: Nabu – the (wise) Man (the god of planet Mercury) to the west, Nergal – the Lion (planet Mars) to the south, Marduk – the Bull (planet Jupiter) to the east, and Ninurta – the Eagle (planet Saturn) to the north. In the Christian tradition, the four symbols have been related to the four Evangelists: St. Matthew's sign is the Virgin (corresponding to the Man), St. Mark's sign is the Lion, St. Luke's sign is the Bull, and St. John's sign is the Eagle. In Islam, as Ibn 'Arabî stated, the Throne is supported by the same tetramorph: four angels having the form of a man, a lion, an eagle, and a bull.[2] René Guénon, when describing the five *arkân* and the *gammadion*, as a square represented by the four corners (four letters *gamma*) and a central cross, underscored that the *gammadion* was considered to be symbolizing Christ (the central cross) and the four Evangelists, but also Christ in the center of the four creatures of Ezekiel's vision and of the *Apocalypse*, where these four symbolic creatures correspond to the four *Mahârâjas* who are, in the Hindu and Tibetan traditions, the regents of the cardinal points.[3]

---

[1] *Revelation* 7:1.
[2] Charles-André Gilis, *La doctrine initiatique du pèlerinage*, Les Éditions de l'Œuvre, 1982, pp. 40-41.
[3] *Symboles fondamentaux*, p. 299. Guénon also said that the tetramorph represents the synthesis of the four elementary powers (*Le symbolisme de la croix*, p. 64).

This whole picture raises the question: are the four winds, the four horses, or the four creatures, guardians of the cardinal points or of the corners of the earth? In fact, there is a dual symbolism, which we could describe using the two forms of the *tetra-gammadion*: the first, also called *gammadia* or the "cross voided throughout,"[1] represents a cross composed of four Greek letters *gamma* with the summits pointing to the center, where the inner void in the shape of a cross symbolizes Jesus Christ and the four *gammas* designate the four Evangelists[2]; the second is the one presented above with the summits of the four *gammas* marking the corners of a square. In the first case, the four branches of the cross are oriented towards the cardinal points; in the second case, the four corners are in focus.[3] With regard to the four *arkân* symbolizing in the Alchemy the four elements, Guénon said: "This assimilation of the elements to the four angles of a square is also naturally related to the correspondence that exists between the same elements and the cardinal points."[4]

Because the earth was symbolically seen as a square or a cube, being the result of the Center manifested as temple,[5] the four corners were often used to represent the whole manifestation, replacing the cardinal points and creating a picture that not only had in mind the corners of a temple – image of the Cosmos, but was also related to the center as the intersection of the diagonals connecting the four corners.[6] Therefore, at the end of this world, the four corners have an

---

[1] See the first chapter.
[2] Guénon, *Le symbolisme de la croix*, p. 71, *Symboles fondamentaux*, p. 298.
[3] In the Christian tradition, the most common image is the one having the Evangelists in the corners or occupying the four quarters (see for example *The Book of Kells*); but also the other form was in use (see for instance *Codex Aureus* of Speyer Cathedral).
[4] *Symboles fondamentaux*, p. 283. There is a strong correlation between the alchemical symbolism and the architectural one (*ibid.* pp. 284-285, 291).
[5] This did not mean that the medieval and ancient people believed the earth was flat.
[6] We remember from the preceding chapter the St. Andrew's cross made with ashes at the dedication of the temple.

important symbolic double role, to express both the dissolution of this cycle (the operation of the four horsemen) and the reassembly into the Center.[1] "Moreover the word of the Lord came unto me, saying, Also, thou son of man, thus saith the Lord God unto the land of Israel; An end, the end is come *upon the four corners of the land*.[2] Now is the end come upon thee, and I will send mine anger upon thee, and will judge thee according to thy ways, and will recompense upon thee all thine abominations."[3] "And he shall set up an ensign for the nations, and shall assemble the outcasts of Israel, and gather together the dispersed of Judah from *the four corners of the earth*."[4]

The combination of the four cardinal points with the four corners is often called the Wind Rose, which tries to integrate the two symbolic meanings into one. However, there is an essential difference between the two representations, since the first one (that is, the cross uniting the four cardinal points and inscribed in a circle or a square) is primarily the symbol of the immutable Center, while the second (the diagonals connecting the four corners) describes the Activity of the Center,[5] in the same way as the *swastika*, which is a sign of the Pole, signifies the Principle's action upon the world.[6] Guénon specified that the *swastika* suggests the rotation of the universal vortex and substitutes the circle with four cardinal tangents, and this is related to the ritual circumambulation,[7] which, in the Hindu

---

[1] If "measurement" means manifestation, then we understand the Hindu statement "the Metres are the Quarters" (Ananda K. Coomaraswamy, *Spiritual Authority and Temporal Power in the Indian Theory of Government*, Munshiram Manoharlal Publishers, 1978, p. 83).

[2] Our italic.

[3] *Ezekiel* 7:1-3.

[4] *Isaiah* 11:12. Our italic.

[5] Only the corners can be physically marked (by the "angle–stones"), while the four cardinal points are subtly indicated by the temple's orientation. On the other hand, in the case of the octagon, the corners are converted into elements of the subtle domain.

[6] Guénon, *Le symbolisme de la croix*, p. 72.

[7] Concerning the meaning of circumambulation see our work *The Everlasting Sacred Kernel*, Rose-Cross Books, 2001, p. 64. The ritual of circumambulation is essentially connected to the temple.

tradition, in the case of a solar orientation (having the center always to the right), follows the path north-east-south-west (*pradakshinâ*).¹ Therefore, Guénon explained, in the Vedic symbolism, the *déva-loka* gate is placed at north-east (in a "corner" and not in a cardinal point) and the *pitri-loka* gate at south-west, implying the rotation of the cycle.² Thus, when the Masonic operative ritual requires the first stone to be placed in the north-east angle, this is not only to mark the corners, but also to suggest the solar rotation around the center, since, as Guénon said, the stones of the other angles were placed following the apparent route of the sun, that is, in the south-east, south-west, and north-west.³

"The round roof of the Ming Tang is sustained by eight pillars, which are supported by a square base, like the earth, and to realize this quadrature of the circle, which goes from the heavenly unity of the vault to the quaternary of the terrestrial elements, it must pass through the octagon, which refers to the intermediary world of the eight directions, of the eight gates and eight winds."⁴ René Guénon, quoting this text,⁵ considered that "the square is associated to the four cardinal points and their various traditional meanings. To obtain the octagonal form, it is necessary to take into account, between the four cardinal points, the four intermediary points, which together compose the assembly of the eight directions that are what different traditions call the 'eight winds.' ... In the Vedic ternary of the gods governing the Three Worlds, *Agni*, *Vâyu*, and *Aditya*, *Vâyu* [the Wind] corresponds to the intermediary world."⁶ Guénon compared the term "Wind Rose" with the Rosicrucian's symbols *Rosa Mundi* and *Rota Mundi*, where *Rosa Mundi* had eight rays corresponding to the elements and the sensible qualities; indeed, Aristotle presented the four corporeal

---

¹ Guénon, *La Grande Triade*, p. 71.
² *Symboles fondamentaux*, p. 241.
³ *Symboles fondamentaux*, p. 280.
⁴ Benoist, *Art du Monde*, p. 90.
⁵ *Symboles fondamentaux*, p. 276.
⁶ *Symboles fondamentaux*, p. 275.

elements placed in the cardinal points, and the sensible qualities in the four corners or intermediary points.[1]

In the Far-Eastern tradition, the center itself had sometimes the octagonal form; Luc Benoist mentioned a city with eight gates, illustrating the eight winds, capable of attracting the best cosmic influences.[2] A common form for the *Ming Tang* was the octagon, a central room and eight around it,[3] and the eight primary *guas* or trigrams were too placed in an octagonal form.[4] Guénon also described an Assyrian image of the sun with eight rays, and he related the number eight to the Christian symbolism of *Sol Justitiae*; it is interesting that the solar god situated near this image had in one hand "a disc and a rod, which are the conventional representations of the measuring ruler and the rod of justice," and Guénon recalled the relation between "measure" and the "solar rays,"[5] which should be in addition connected to the importance of the right measurements presented above with regard to the center. One of the main significances of the number 8, René Guénon underscored, is that of "justice" and "equilibrium," directly attached to the idea of Center.[6] Furthermore, René Guénon presented the *betyl* of Kermaria[7] (which was considered the *omphalos*, that is, the center of Brittany or Armorica) having on one face a *swastika* and on another face a figure of eight rays circumscribed by a square (instead of a circle, like in the case of the wheel with eight spokes).[8] This last figure brings us back to

---

[1] Guénon, *Études sur l'hindouisme*, p. 50. However, as Guénon specified, there are symbols involving eight angels, and the winds could be considered "messengers" of God (that is, "angels") (*Psalms* 104:4), which means that the octagon could be more than a representation of the intermediary world (*Symboles fondamentaux*, p. 277).
[2] *Art du Monde*, p. 91.
[3] Marcel Granet, *La pensée chinoise*, Albin Michel, 1968, p. 151, and our work *About the Yi Jing*, Rose-Cross Books, 2006, pp. 51-52.
[4] See also Guénon, *La Grande Triade*, p. 199, where Guénon mentioned the *Dharma-chakra* as well, as a wheel with eight spokes.
[5] *Symboles fondamentaux*, p. 362.
[6] *Écrits*, p. 93, *Symboles fondamentaux*, pp. 101, 362.
[7] Breton *kêr* = house; *Maria* = the Virgin Mary.
[8] *Écrits*, p. 92.

The vision of the Center 145

the temple as center, where the corners and the cardinal points play an essential role.

Guénon noticed the importance of the square with the eight rays in relation to the Templars and Masonry, since this figure was discovered, in more than one exemplar, in the graffito from the donjon of Chinon.[1] Louis Charbonneau-Lassay, who thoroughly studied such symbols, including these from Chinon, concluded that we can accept as accurate the local tradition which attributed the engraved images to a leader of the Order of the Temple, who was once captive in the donjon of Chinon.[2] The "Wind Rose" is present at Chinon not only inscribed in a square, but also on a Knight Templar's shield as a coat of arms, and Charbonneau-Lassay gave two other examples where this "heraldic sign" appeared in relation to the Templars.[3] Besides this symbol, he discovered at Chinon another important one, called the "triple–enceinte," that was carved by the same Templar.[4]

There is no doubt that the "triple-enceinte," likewise the "Wind Rose," is directly related to the idea of center, and in particular to the idea of the temple as center. The "triple-enceinte" represents three concentric squares and a cross formed by four lines that link the centers of the four sides of the inner square with the centers of the correspondent sides of the other two squares.[5] René Guénon studied the symbolism of the "triple-enceinte," in correlation to the "Wind Rose" from Chinon (the square containing 8 rays), and considered that the

---

[1] *Écrits*, p. 92.
[2] Louis Charbonneau-Lassay, *Le Coeur rayonnant du donjon de Chinon*, Archè, 1975, p. 33.
[3] *Le Coeur rayonnant*, p. 16. The two examples are: a funeral statue of a Templar (the Commandery of the Temple at Roche-en-Cloué) and a carved stone (Commandery of the Temple at Mauléon). Guénon affirmed: "this sign must have played an important role in the symbolism of the Templars"; he added: "this figure survived in the modern Masonry" (*Écrits*, p. 93).
[4] Louis Charbonneau-Lassay, *L'ésotérisme de quelques symboles géométriques chrétiens*, Éditions Traditionnelles, 1985, p. 16.
[5] Charbonneau-Lassay mentioned also another graffito found on a stone of a chapel discovered at Seuilly, where the "triple-enceinte" has an octagonal form (*ibid.* p. 17).

three enceintes represent the three principal degrees of initiation,[1] but, at the same time, they could signify three states of the being or the Three Worlds of the Hindu tradition.[2]

The "triple-enceinte" could have other essential meanings. The plan of Solomon's Temple was tripartite: the porch (*ulam*), the main sanctuary (*hekam*), and the Holy of Holies (*debir*), representing the same spiritual gradation as the "triple-enceinte," which makes this a symbol of the temple as center. René Guénon understood very well the significance, since he presented this figure in a modified form as well: if the middle square is rotated by 90°, its corners will mark the centers of the four sides of the outside square, while the corners of the inner square (unmoved) will touch the centers of the four sides of the middle square just rotated; at the same time, the cross will rotate with the middle square and will connect the four corners of the inner square with the four corners of the outside square. In this way, all the eight directions are marked on the outside square: the four cardinal points (marked by the corners of the middle square) and the four "corner–stones" (marked by the four branches of the cross). This new diagram, Guénon said, is the one used by the ancient astrologers to inscribe the Zodiacal signs and symbolized Heavenly Jerusalem,[3] that is, the supreme Center.

Solomon's Temple was the image of the supreme Center in the same way the "foundation–stone" was the direct projection of the "keystone," since each secondary center was viewed not so much as a "corner–stone" but as a "foundation–stone," which implied that it represented a different situation or aspect of the same Center.[4] Therefore, according to the Jewish tradition, the "foundation–stone" (*Even ha-Shethiyah*) was the

---

[1] This division is the most regular and fundamental one, as it is seen in Masonry, for example. Guénon added that in some Masonic high degrees systems the degrees were described precisely as enceintes.
[2] *Symboles fondamentaux*, pp. 99-102.
[3] *Symboles fondamentaux*, p. 104.
[4] "Behold, I lay in Sion for a foundation a stone, a tried stone, a precious angle stone, a sure foundation" (*Isaiah* 28:16).

rock marking the Center of the World: symbolically, this rock was the altar used by Adam and Noah to offer sacrifices to God; it was Abraham's altar, where he was willing to sacrifice his son Isaac; it was the stone upon which Jacob dreamt about the angels ascending and descending on a ladder; it represented Mount Moriah upon which Solomon built his Temple.[1] Moreover, this rock, this "foundation–stone" was the center from which the world was produced as explication. We read in the *Zohar*: "The world did not come into being until God took a certain stone, which is called the 'foundation–stone' (*Even ha-Shethiyah*), and cast it into the abyss so that it held fast there, and from it the world was planted. This is the central point of the universe, and on this point stands the Holy of Holies. This is the stone referred to in the verses, 'Who laid the corner–stone thereof,'[2] 'the stone of testing, the precious corner–stone,'[3] and 'the stone that the builders despise became the head of the corner.'[4]"[5] In the *Midrash Tanhuma*, the center is specified as a symbolic hierarchy of the spiritual Pole's successive approximations: "As the navel is set in the center of the human body, so is the land of Israel the navel of the world ... situated in the center of the world, and Jerusalem in the center of the land of Israel, and the sanctuary in the center of Jerusalem, and the holy place in the center of the sanctuary, and the ark in the center of the holy place, and the Foundation–Stone before the holy place, because from it the world was founded."

For the Islamic tradition, this "foundation–stone" is the "Rock," *As-Sakhrah*, which came from Paradise, and from where Muhammad the Prophet ascended to heaven. The *Qubbat as-Sakhrah*, the Dome of the Rock, built in this very holy place,

---

[1] A Masonic tradition regarded the "foundation-stone" of Solomon's Temple as a double cube stone with the sides engraved by Noah, Abraham, Moses, Joshua, and Hiram Abif. The name of the main temple of Babylon was *E-temen-an-ki*, "Temple of the Foundation of Heaven and Earth."
[2] *Job* 38:6.
[3] *Isaiah* 28:16.
[4] *Psalms* 118:22.
[5] *The Zohar*, The Rebecca Bennet Publications, 1956, vol. II, p. 339, *Vayehi* 231a.

was considered the "image of the spiritual center of the world,"[1] and it represented a dome (*qubba*) crowning an octagon, a geometrical form Burckhardt considered directly related to the form of the Byzantine churches,[2] but which, more importantly, is a plain illustration of the octagon's symbolism presented above. The *Qubbat as-Sakhrah* had four gates: "four portals open out onto the four cardinal points, thus placing the building symbolically at the center of the world."[3] And Burckhardt added: "Thus, the very plan of the sanctuary expresses the synthesis of the circle and the square, movement and repose, time and space, and this synthesis is already expressed, in the most striking fashion, by the exterior shape of the building in which the 'celestial' sphere of the dome marries the 'earthly' crystal of the octagon."[4] In the Islamic tradition as well, the octagon was related to the Vision of the Center.

However, we have to keep in mind that no people or nation could claim to be the owner of the Rock itself, which, being the symbol of the Center as temple, is in fact the owner of peoples and nations.

---

[1] Titus Burckhardt, *Art of Islam*, World of Islam Festival Publishing, 1976, p. 12.
[2] *Art of Islam*, p. 10.
[3] *Art of Islam*, p. 12.
[4] *Art of Islam*, p. 12. Burckhardt noted that usually the octagon is an intermediary between the dome and the cubic base, having a different spiritual significance.

VI

SOLOMON'S TEMPLE

The Christian tradition had in view as model for its temple both Heavenly Jerusalem and Ezekiel's Temple, yet Solomon's Temple remained an essential symbol of the Center.[1] There were subtle reasons why the Temple of Solomon continued to play a mythical function in the Christian tradition and especially in Christian esotericism, despite the fact that the area of the Temple Mount was related to the "abomination of desolation" and, after the destruction of the Temple, it was left in ruins – a desolated place witnessing the Justice of God.[2]

---

[1] For the Judaic tradition, Solomon's Temple was unique as the center is: "As God is one, his temple also should be one. In the next place, he does not permit those who desire to perform sacrifices in their own houses to do so, but he orders all men to rise up, even from the furthest boundaries of the earth, and to come to this temple" (Philo, *De Specialibus Legibus*, I, 67-68). Since the Temple of Jerusalem was the place where the *Shekinah* resided constantly, the sacrifices belonging to the public ritual had to be offered in this unique location (Guénon, *Écrits*, pp. 110-111).

[2] "And arms shall stand on his part, and they shall pollute the sanctuary of strength, and shall take away the daily sacrifice, and they shall place the abomination that maketh desolate" (*Daniel* 11:31); "And Jesus said unto them, See ye not all these things? verily I say unto you, There shall not be left here one stone upon another, that shall not be thrown down" (*Matthew* 24:2); "When ye therefore shall see the abomination of desolation, spoken of by Daniel the prophet, stand in the holy place …" (*Matthew* 24:15); "And when ye shall see Jerusalem compassed with armies, then know that the desolation thereof is nigh" (*Luke* 21:20); "But when ye shall see the abomination of desolation, spoken of by Daniel the prophet, standing where it ought not, (let him that readeth understand,) then let them that be in Judaea flee to the mountains" (*Mark* 13:14).

René Guénon considered the 6th Century B.C. a temporal barrier that marked the beginning of the "historical" period and beyond which there was an inaccessible "mythical" past (for the common researchers). Considerable changes depicted this barrier; in China, the tradition was parted in two: Daoism and Confucianism; in Persia, Zoroaster readapted Mazdeism; in India, Buddhism emerged; in Europe, the "historical" period of Rome commenced; and it was the rise of Pythagorism.[1] This temporal barrier, Guénon said, was also related to another event: the Babylonian captivity, during which the Jews lost a part of their heritage and had to rewrite their sacred books with new characters. Guénon mentioned another critical epoch, which represented a revival of the Occidental world, when the Christian tradition came to light and the Jewish people's dispersal (Greek διασπορά) occurred.[2] For sure, these types of "temporal landmarks" are directly related to the doctrine of the cosmic cycles,[3] which the Judaic tradition was well aware of: "Thou, O king, art a king of kings: for the God of heaven hath given thee a kingdom, power, and strength, and glory. ... Thou art this head of gold. And after thee shall arise another kingdom inferior to thee, and another third kingdom of brass, which shall bear rule over all the earth. And the fourth kingdom shall be strong as iron. ... And in the days of these kings shall the God of heaven set up a kingdom, which shall never be destroyed: and the kingdom shall not be left to other people, but it shall break in pieces and consume all these kingdoms, and it shall stand for ever."[4]

Daniel's prophecy and translation of the king's dream are, in fact, an overview of the doctrine of the cosmic cycles applied to our present *Manvantara*, with its four Ages,[5] including the end of times, when the four winds of the heaven will "consume all

---

[1] *La crise du monde moderne*, pp. 20-24.

[2] *La crise*, p. 27.

[3] As Ibn 'Arabî was saying, "all the ways are circles; there is no straight line" (*Journey to the Lord of Power*, Inner Traditions International, 1989, p. 40).

[4] *Daniel* 2:37-44.

[5] See Gaston Georgel, *Les Quatre Âges de l'Humanité*, Archè, 1976, p.120.

these kingdoms" and everything will be reintegrated in the everlasting Center. Nonetheless, the Judaic prophetic tradition is concerned not so much with the whole cycle, but with the secondary ones. As Guénon said, "it seems that the Biblical flood corresponds directly to the cataclysm in which Atlantis disappeared, and, consequently, it must not be identified with the flood of *Satyavrata*, which, in conformity with the Hindu tradition that derived directly from the Primordial Tradition, preceded the beginning of our *Manvantara*."[1]

Atlantis represented only one of the *Manvantara*'s five "Grand Years"[2] and was situated half in the *Trêtâ-yuga* (the "Silver Age") (Plato's "divine period") and half in the *Dwâpara-yuga* (the "Bronze Age") (Plato's "human period"). However, due to the analogy that exists between a principal cycle and the secondary ones (and also between the secondary cycles themselves), it is always possible to apply the various data and interpretations to different levels,[3] and hence Zechariah's prophecies, for example, could envision the end of a secondary cycle and not of the entire *Manvantara*.[4]

The Judaic prophetic tradition was very closely related to Solomon's Temple and, in fact, the Temple appeared as the primeval symbol for the doctrine of the cosmic cycles applied to the post-Atlantis period. It appears that the description of the Jews as "chosen people" has a concealed meaning,[5] related to the fact that the Jewish people was preserved during this entire post-Atlantis period[6] seemingly to illustrate the cyclic

---

[1] *Formes traditionnelles et cycles cosmiques*, p. 49.
[2] Guénon, *Formes traditionnelles et cycles cosmiques*, p. 48.
[3] *Formes traditionnelles et cycles cosmiques*, pp. 14, 49.
[4] "Here have been, and will be again, many destructions of mankind arising out of many causes; the greatest have been brought about by the agencies of fire and water, and other lesser ones by innumerable other causes" (Plato, *Timaeus*, 22b).
[5] From a pure esoteric perspective, Guénon pointed out that the initiates are the real "chosen people" and Israel symbolizes the "assembly of the initiates" (*Aperçus sur l'ésotérisme chrétien*, p. 85).
[6] In a letter to Goffredo Pistoni (July 24, 1949), René Guénon wrote: "The question regarding the relation between the Judaism and Christianity is much

evolvement of the world until its very end.¹ For the same reason, in Masonry, the importance of Solomon's Temple is related to the tradition that considers the "Noachite" origin of Free-Masonry,² and the existence of many Judaic elements in the Masonic rituals is, at least in part, motivated by this concealed meaning.³

The Atlantis' influence spread into the world far before its destruction, which facilitated its heritage to be well received, as the myths of the Toltec, Aztec, and Maya civilizations attested.

---

more complex than how you see it, because this does not explain the persistency of the Judaism till nowadays."

[1] There is a relation with "a dry wind of the high places in the wilderness" (*Jeremiah* 4:11). "But they hearkened not, nor inclined their ear, but walked in the counsels and in the imagination of their evil heart, and went backward, and not forward [which means "downward" with the cycle]. Since the day that your fathers came forth out of the land of Egypt unto this day I have even sent unto you all my servants the prophets, daily rising up early and sending them: Yet they hearkened not unto me, nor inclined their ear, but hardened their neck: they did worse than their fathers" (*Jeremiah* 7:24-26). The Prophet Muhammad presented again and again the example of the prophets from Abraham, the true and orthodox believer, to Jesus, and how they were rejected by the Israelites, an example the Arabs and the others should not follow (with regard to Muhammad). The Prophet made of the punishment of the Jews a fundamental and emblematic teaching: "Then were they smitten with abasement and poverty, and met with wrath from God. That was because they had misbelieved in God's signs and killed the prophets undeservedly; that was for that they were rebellious and had transgressed" (*Qur'an* 2:61); "Verily, those who disbelieve in God's signs, and kill the prophets without right, and kill those from among men, who bid what is just, – to them give the glad tidings of grievous woe!" (3:21). The *Qur'an* gives persistently the Jews as example: "Ask the children of Israel how many a manifest sign we gave to them" (2:211); "Dost thou not look at the crowd of the children of Israel after Moses' time, when they said to a prophet of theirs…" (2:246), and so on.

[2] See Guénon, *Études sur la Franc-Maçonnerie*, I, p. 199. The building of the Ark was a Masonic operation. The Mason's initiatory main task was to build and rebuild the center, to suffer a sacrifice and rise again. Another hint with respect to Atlantis is that Hiram Abif was a "worker in brass" (and so was Azazel in the *Book of Enoch*).

[3] As Guénon said, the Hebraic elements are related to an esoteric side and have nothing to do with the political aspects (*Études sur la Franc-Maçonnerie*, I, p. 276); the wild fantasies regarding the so-called "Masonic-Judaic conspiracy" are well known.

Solomon's Temple 153

It is interesting to note how the "terrible" end of this secondary cycle was kept in the memory of these peoples, who developed, due as well to their incomplete traditional knowledge, an almost obsessive concern with regard to the end of times, which made them stress the importance of the calendar and of the sacrifices. The Egyptian tradition derived also from the Atlantean one,[1] to which we must add some southern influences; the Chaldean tradition was also a beneficiary of the Atlantean heritage, but combined with Hyperborean influences, and the Judaic tradition, which is essentially "Abrahamic," derived from this Chaldean tradition (later on, Moses operated a "revision," when Egyptian elements, concerning especially some traditional sciences, were added).[2] The "antediluvian" origins of Masonry are closely connected to the above considerations and we should recall that the Masonic symbols are an expression of certain traditional sciences belonging to Hermeticism,[3] which is a tradition of Egyptian origin transmitted through a Hellenistic form to Christianity and Islam.[4] In the Islamic tradition, Seyîdna Idris was identified at the same time with Hermes and Enoch, which implies a continuity of Hermeticism beyond the Egyptian sacerdotal depot to the Atlantean tradition[5]; there are mentioned sometimes three distinct Hermes: *Hermes El-Harâmesah* (the Hermes of all Hermes), identified to Seyîdna Idris; *Hermes El-Bâbelî* (the Babylonian Hermes); and *Hermes El-Miçrî* (the Egyptian Hermes), all this suggesting that the Chaldean and Egyptian traditions derived from the Atlantean one.[6]

---

[1] However, we should stress that the Atlantean tradition was not the Primordial Tradition.
[2] Guénon, *Formes traditionnelles et cycles cosmiques*, p. 153.
[3] Guénon noted the similitude between the names of Hermes and Hiram, which even though they do not have the same linguistic origin, are identical with respect to their composition (HRM) (*Formes traditionnelles*, p. 129).
[4] Guénon, *Formes traditionnelles*, p. 120, *Franc-Maçonnerie*, I, p. 17. Nonetheless, Hermeticism does not represent the entire Egyptian tradition, the same way the latter does not represent the entire Atlantean tradition.
[5] Guénon, *Formes traditionnelles*, pp. 133, 142.
[6] Guénon, *Formes traditionnelles*, p. 146.

The Jewish people also received the Atlantis' heritage, mainly through the intermediary of the Egyptians,[1] and the devastation of the "red–race" became a fundamental stone for the Judaic tradition, which translated the doctrine of the cosmic cycles into an alternation of justice (the "stumbling–stone") and mercy (the "keystone"), of punishment and regeneration, illustrating what the Hindu tradition calls the "inspiration" and "expiration" of *Brahma*'s Breath and the fact that the *Shekinah* has two main aspects: Mercy and Justice.[2]

The Judaic tradition interpreted the Atlantis' cataclysm as God's punishment, but, like always, the punishment became inevitable only after God's mercifulness exhausted all its possibilities; for this reason, only when Noah could not find the maximum of "ten righteous men" in the world, did God send the flood[3]; and, in the same way, He punished Sodom only after Abraham could not find ten righteous people.[4] Each time in

---

[1] Guénon, *Formes traditionnelles*, pp. 50, 153. "And Melchizedek king of Salem brought forth bread and wine: and he was the priest of the most high God. And he blessed him, and said, Blessed be Abram of the most high God, possessor of heaven and earth" (*Genesis* 14:18-19). For René Guénon, this event symbolized the connection point between the Judaic tradition and the Primordial Tradition (*Le Roi*, p. 50).

[2] Guénon, *Le Roi*, pp. 25-26. As we said, the Prophet Muhammad made of this punishment of the Jews a fundamental and emblematic teaching: "Then were they smitten with abasement and poverty, and met with wrath from God. That was because they had misbelieved in God's signs and killed the prophets undeservedly; that was for that they were rebellious and had transgressed" (*Qur'an* 2:61). But the punishment was always overcome by God's mercy, and the *Qur'an* recorded the story of Ezra who, passing by the ruins of Jerusalem (destroyed by the Chaldeans), doubted God's power to restore the city and its inhabitants; whereupon God caused him to die and resurrected him after a hundred years ("Or like him who passed by a village, when it was desolate and turned over on its roofs, and said, How will God revive this after its death? And God made him die for a hundred years, then He raised him," *Qur'an* 2:259).

[3] *Zohar*, I, 67b, see also our *The Wrath of Gods*, p. 180.

[4] "And the Lord said, If I find in Sodom fifty righteous within the city, then I will spare all the place for their sakes. ... And he said unto him, Oh let not the Lord be angry, and I will speak: Peradventure there shall thirty be found there. And He said, I will not do it, if I find thirty there. ... And He said, I will not destroy it for ten's sake" (*Genesis* 18:26-32). And later, Jeremiah said the same

Solomon's Temple 155

these cases the cycle, principal or secondary, decayed so much that the end could not be avoided anymore, and this idea the Judaic tradition carried with itself all along the post-Atlantis period, with the specification that a traditional society or a cycle is finished when its center is obliterated.[1] Therefore, we find references to Shiloh as an example of a destroyed center: "But go ye now unto my place which was in Shiloh, where I set my name at the first, and see what I did to it for the wickedness of my people Israel"[2]; "Then will I make this house like Shiloh, and will make this city a curse to all the nations of the earth."[3] There is no clear reference regarding the destruction of Shiloh; on the contrary, the Philistines, because they abducted the Ark

---

thing in connection to Jerusalem: "Run ye to and fro through the streets of Jerusalem, and see now, and know, and seek in the broad places thereof, if ye can find a man, if there be any that executeth judgment, that seeketh the truth; and I will pardon it" (5:1). We may note that righteousness is symbolized by "judge [make justice to] the fatherless, plead for the widow" (*Isaiah* 1:17, 1:23), a formula adopted by the Templars and Masonry; Isaiah calls the "new" Jerusalem (after the punishment of the sinners) "the City of righteousness" (1:26). "Behold, the days come, saith the Lord, that I will raise unto David a righteous Branch, and a King shall reign and prosper, and shall execute judgment and justice in the earth. In his days Judah shall be saved, and Israel shall dwell safely: and this is his name whereby he shall be called, The Lord our Righteousness" (*Jeremiah* 23:5-6). And Malachi: "But unto you that fear my name shall the Sun of righteousness arise with healing in his wings" (4:2).

[1] Since the Jewish prophets themselves described the lack of righteousness, St. Paul's words are nothing else than a follow up: "What shall we say then? That the Gentiles, which followed not after righteousness, have attained to righteousness, even the righteousness which is of faith. But Israel, which followed after the law of righteousness, hath not attained to the law of righteousness. Wherefore? Because they sought it not by faith, but as it were by the works of the law. For they stumbled at that stumbling–stone; As it is written, Behold, I lay in Sion a stumbling–stone and rock of offence: and whosoever believeth on him shall not be ashamed" (*Romans* 9:30-33); to be noted the "stumbling–stone."

[2] *Jeremiah* 7:12.

[3] *Jeremiah* 26:6, 26:9. Before Jerusalem, Shiloh represented the spiritual center where the Tabernacle was placed: "And the whole congregation of the children of Israel assembled together at Shiloh, and set up the tabernacle of the congregation there" (*Joshua* 18:1); see also *1 Kings* 1.

of the Covenant,[1] suffered God's retribution,[2] while in other cases the various nations that attacked Israel were considered instruments of God's Justice, even though they were also punished in the end. However, this event marked the end of the spiritual center, Shiloh, since: "The glory [*Shekinah*] is departed from Israel: for the ark of God is taken,"[3] which means the place was emptied of the spiritual influences; moreover, it signified that a secondary cycle ended and a new one would begin having Jerusalem as center, and thus, for Jeremiah, it represented a model for the new center and the new cycle, which could also suffer God's Rigor and the departure of the *Shekinah*.

The construction of Solomon's Temple indicated the foundation of a new secondary center, representing a combination of the "center as temple" and the "temple as center," since Mount Moriah, and especially the Rock, was considered to be Abraham's altar and also David's altar. We may say that if the construction and destruction of the Temple illustrated the evolvement of the cycles, the immutable Rock symbolized the absolute Principle, the supreme Center, even though for the Judaic tradition the devastation of the Temple sometimes meant the departure of the *Shekinah*, like in the case of Shiloh.[4]

---

[1] *1 Kings* 4, *1 Kings* 5.
[2] *1 Kings* 5:6-9.
[3] *1 Kings* 4:22.
[4] From the Temple's perspective Mount Moriah is equivalent to Mount Sion and represents the Center as temple ("Nevertheless David took the strong hold of Sion: the same is the city of David," *2 Kings* 5:7; "So David went and brought up the ark of God from the house of Obededom into the city of David with gladness," *2 Kings* 6:12; "Then Solomon assembled the elders of Israel, and all the heads of the tribes, the chief of the fathers of the children of Israel, unto king Solomon in Jerusalem, that they might bring up the ark of the covenant of the Lord out of the city of David, which is Sion," *3 Kings* 8:1). In Masonry, the first Lodge was held in a valley where peace, truth, and union reigned, in the midst of three mountains: Moriah, Sinai and Heredom.

Solomon's Temple

It must be stressed that, as a secondary center, the Temple had a heavenly "prototype"[1] and in itself has no "historical" origin, since the manifestation of the Center as temple (*Shekinah*) means a "non-human" origin, the same way the initiation or Masonry has no historical origin for the simple reason that the real origin is situated in a world where the conditions of time and space that define the historical facts don't apply.[2] From a Judaic perspective, Solomon's Temple was not just the First Temple, but the only Temple, the perfect center,[3] everything else built afterwards being incomplete and increasingly lame. Since the Second Temple did not explicitly contain the Ark,[4] and other sacred things were missing, it was viewed sometimes as incomplete and, even more, without the *Shekinah*, because the glory of God did not return. On the other hand, there is a strong Jewish belief that the *Shekinah* never left the Temple Mount, which remained a holy place, that is, which still could be considered a secondary center.

---

[1] "Thou hast commanded me to build a temple upon thy holy mount ... a resemblance of the holy tabernacle, which thou hast prepared from the beginning" (*Wisdom of Solomon* 9:8).

[2] Guénon, *Aperçus sur l'initiation*, p. 58. "Jerusalem is understood historically of that earthly city whither pilgrims journey; allegorically, of the Church militant; tropologically, of every faithful soul; anagogically, of the celestial Jerusalem, which is our Country" (Durantis 10). We may remember that Solomon's Temple is of divine inspiration and, in fact, God is the architect: "Where wast thou when I laid the foundations of the earth? declare, if thou hast understanding. Who hath laid the measures thereof, if thou knowest? or who hath stretched the line upon it? Whereupon are the foundations thereof fastened? or who laid the corner stone thereof?" (*Job* 38:4-6). In Masonry, Solomon is himself an architect.

[3] Solomon's Temple was a "condenser" of the spiritual influences (Guénon, *L'Erreur*, p. 58).

[4] The last time the Ark was mentioned was during Josiah's reign, before the Babylonian captivity: "And said unto the Levites that taught all Israel, which were holy unto the Lord, Put the holy ark in the house which Solomon the son of David king of Israel did build; it shall not be a burden upon your shoulders: serve now the Lord your God, and his people Israel" (*2 Chronicles* 35:3). There are certain opinions that assume the Ark was buried in a secret chamber under the temple.

The very first Temple for the Judaic tradition was the Tabernacle built by Bezaleel and Aholiab[1] in accord with God's instructions given to Moses, and it represented the center.[2] However, even though the Tabernacle was not destroyed, meaning that the *Shekinah* did not leave the Jewish people, we may notice the cycles occurring during the long "journey" to Jerusalem. The first "fall" took place just before the construction of the Tabernacle, when the Jews built the "golden calf" and Moses "cast the tablets out of his hands, and broke them beneath the mount,"[3] which means breaking the bond with the Principle[4]; a new cycle starts with the new "tablets of stone"[5] and the Tabernacle is built, which, in order to be the temple as center, had to be anointed, and so had to involve the

---

[1] *Exodus* 31, 36; they were architects filled with the Spirit of God.

[2] "And let them make me a sanctuary; that I may dwell among them" (*Exodus* 25:8). "And thou shalt put the mercy seat above upon the ark; and in the ark thou shalt put the testimony that I shall give thee. And there I will meet with thee, and I will commune with thee from above the mercy seat, from between the two Cherubim which are upon the ark of the testimony, of all things which I will give thee in commandment unto the children of Israel" (*Exodus* 25:21-22).

[3] *Exodus* 32.

[4] Léo Schaya compared the destruction of the two Temples to the breaking of the tablets of stone (*Le Temple de Salomon*, Études Traditionnelles, no. 432-433, 1972).

[5] *Exodus* 34. The Islamic tradition affirms that Moses received nine tablets, seven made of stone (which were revealed to the Israelites) and two made of light (which Moses was forbidden to reveal); it was Christ's mission to reveal these two (*Moisis doctrina velat quod Christi doctrina revelat*, the Abbot Suger said). The Prophet Muhammad had a central position, revealing more than Moses and veiling more than Christ; Islam is therefore the "religion of the middle" (see Titus Burckhardt's commentaries in *De la Thora, de l'Evangile et du Qorân*, Études Traditionnelles, no. 224-225, 1938). Schuon, comparing Christ to Muhammad, said that the Prophet appeared mainly under a human aspect not because of some individual impotence but because Islam's reason to be was different from that of Christianity (Frithjof Schuon, *Du Christ et du Prophète*, Études Traditionnelles, no. 224-225, 1938); however, René Guénon explained that the conceptions of Prophet and *Avatâra* proceed inversely one from another, the second one illustrating the Principle manifesting itself, while the first one illustrates the "support" of this manifestation (*Aperçus sur l'ésotérisme islamique*, pp. 59-60) (Christ is the *Avatâra*, Muhammad the Prophet).

Solomon's Temple   159

priests, who performed rites that brought the spiritual influences, and therefore only after the anointment the *Shekinah* came into the Tabernacle.[1]

The following "fall" occurred when the Jews "murmured against Moses and against Aaron: and the whole congregation said unto them, Would God that we had died in the land of Egypt! Or would God we had died in this wilderness!"[2] This transgression God punished with the forty years of wandering in the wilderness,[3] and Moses and Aaron suffered the same punishment since they could not prevent a second "fall."[4] It is interesting to note how at the end of this minor cycle the center comes into view: it is marked by the Stone from which Moses drew water by striking it with his rod.[5]

---

[1] "And thou shalt take the anointing oil, and anoint the tabernacle, and all that is therein, and shalt hallow it, and all the vessels thereof: and it shall be holy. ... And thou shalt put upon Aaron the holy garments, and anoint him, and sanctify him; that he may minister unto me in the priest's office. ... for their anointing shall surely be an everlasting priesthood throughout their generations" (*Exodus* 40:9-15). "Then a cloud covered the tent of the congregation, and the glory of the Lord filled the tabernacle" (*Exodus* 40:34).

[2] *Numbers* 14:2.

[3] *Numbers* 14:33. "(Their Lord) said: For this the land will surely be forbidden them for forty years that they will wander in the earth, bewildered" (*Qur'an* 5:26); the Islamic tradition considers that for forty years the Jews wandered in a circle, each day returning to the same place (which is an illustration of the cycle). Guénon noted the symbolism of the center as "the point around which it is not possible to wander or to err" (*Franc-Maçonnerie*, II, p. 181), which means from the Islamic perspective that the Israelites did not really err, but rotated in the same manner the angels rotate around the Throne.

[4] "And the Lord spake unto Moses and Aaron, Because ye believed me not, to sanctify me in the eyes of the children of Israel, therefore ye shall not bring this congregation into the land which I have given them" (*Numbers* 20:12). See also *Numbers* 27:12-14. We must remember the symbolism of the cosmic cycle, similar in all various traditions, where the Solar Hero or the King is transformed during the descent of the cycle into the Dragon; hence Moses' fate and Solomon's behaviour at the end of his reign.

[5] Numbers 20:11. "This is the water of Meribah; because the children of Israel strove with the Lord, and he was sanctified in them" (Numbers 20:13); we recall the description of the Temple of Jerusalem with the water of life. There is another end of a minor cycle alluding to the center, when Joshua, just before dying, "wrote these words in the book of the law of God, and took a

The destruction of Solomon's Temple illustrating the end of a cycle is anticipated by these various minor cycles; moreover, the role of the foreign armies as God's instruments in carrying on the apocalyptic punishment is exemplified in advance: "And the children of Israel did evil in the sight of the Lord, and served Baalim ... And the anger of the Lord was hot against Israel, and he delivered them into the hands of spoilers that spoiled them, and he sold them into the hands of their enemies round about, so that they could not any longer stand before their enemies. ... And the anger of the Lord was hot against Israel; and he said, Because that this people hath transgressed my covenant which I commanded their fathers, and have not hearkened unto my voice; I also will not henceforth drive out any from before them of the nations which Joshua left when he died: That through them I may prove Israel, whether they will keep the way of the Lord to walk therein, as their fathers did keep it, or not."[1]

David and Solomon's reigns represented another cycle, when the Temple was built, in accordance with God's commands and instruction, and the secondary spiritual center was established in Jerusalem: "God gave Solomon wisdom and very great insight, and a breadth of understanding as measureless as the sand on the seashore. Solomon's wisdom was greater than the wisdom of all the men of the East, and greater than all the wisdom of Egypt."[2] "Solomon sent back this message to Hiram: 'You know that because of the wars waged against my father David from all sides, he could not build a temple for the worship of the Name of the Lord his God until the Lord put his enemies under his feet. But now the Lord my God has given me rest on every side, and there is no adversary or disaster. I intend, therefore, to build a temple for the Name of the Lord my God. So give orders that cedars of Lebanon be cut for me.' In this way Hiram kept Solomon

---

great stone, and set it up there under an oak, that was by the sanctuary of the Lord" (*Joshua* 24:26).
[1] *Judges* 2:11, 2:14, 2:20-22.
[2] *3 Kings* 4:29-30.

Solomon's Temple

supplied with all the cedar and pine logs he wanted. King Solomon conscripted labourers from all Israel – thirty thousand men. Adoniram was in charge of the forced labour. The craftsmen of Solomon and Hiram and the men of Gebal cut and prepared the timber and stone for the building of the temple."[1] "Solomon sent this message to Hiram king of Tyre: 'Send me, therefore, a man skilled to work in gold and silver, bronze and iron, and in purple, crimson and blue yarn, and experienced in the art of engraving, to work in Judah and Jerusalem with my skilled craftsmen, whom my father David provided.' And Hiram added: 'I am sending you Hiram-Abif, a man of great skill, whose mother was from Dan and whose father was from Tyre.'"[2] "In building the temple, only blocks dressed at the quarry were used, and no hammer, chisel or any other iron tool was heard at the temple site while it was being built. The word of the Lord came to Solomon: 'As for this temple you are building, if you follow my decrees, carry out my regulations and keep all my commands and obey them, I will fulfill through you the promise I gave to David your father. And I will live among the Israelites and will not abandon my people Israel.' So Solomon built the temple and completed it. He prepared the inner sanctuary within the temple to set the ark of the covenant of the Lord there. Solomon covered the inside of the temple with pure gold, and he extended gold chains across the front of the inner sanctuary, which was overlaid with gold. In the inner sanctuary he made a pair of cherubim of olive wood, each ten cubits high. He placed the cherubim inside the innermost room of the temple, with their wings spread out. The wing of one cherub touched one wall, while the wing of the other touched the other wall, and their wings touched each other in the middle of the room. He overlaid the cherubim with gold."[3] "He cast two bronze pillars; the height of one pillar was eighteen cubits, and a cord twelve cubits long gave the measurements of its girth; so also was the second pillar. He also

---

[1] *3 Kings* 5:2-18.
[2] *2 Chronicles* 2:3-14.
[3] *3 Kings* 6:7-28.

made two capitals of cast bronze to set on the tops of the pillars; each capital was five cubits high. ... He erected the pillars at the portico of the temple. The pillar to the south he named Jakin and the one to the north Boaz. And so the work on the pillars was completed."[1]

The Biblical narrative regarding Solomon's Temple hides a rich meaning, beyond the particular traditional form to which it belongs. First of all, we note the process of Jews' "stabilization" or "coagulation" when their spiritual center, the Jerusalem Temple, was raised. The Masonic work (in its operative sense) was not typical for nomads, therefore Solomon had to call foreign builders to erect the Temple[2]; in the same way, the construction of Kaaba was conducted by foreign craftsmen. Solomon's Temple, made of wood and stone, was covered with precious stones and gold, being symbolically equivalent to Heavenly Jerusalem; its shape was parallelepiped-like, and the Holy of Holies was cubic.[3] The two pillars made of bronze

---

[1] *3 Kings* 7:15-22. The Islamic tradition, describing how Solomon used *djinns* to build the edifice, mentions that he asked God to conceal his death till the *djinns* finished building the Temple: "And when We decreed death for him, nothing showed his death to them save a creeping creature of the earth which gnawed away his staff. And when he fell the djinns saw clearly how, if they had known the Unseen, they would not have continued in despised toil" (*Qur'an* 34:14). There are other tales about Solomon: how he travelled using a flying carpet of green silk; or that he spoke the language of birds: "And Solomon was David's heir. And he said: O mankind! Lo! we have been taught the language of birds" (27:16) (Tristan also knew the language of birds, see Joseph Bédier, *The Romance of Tristan and Iseult*, Vintage Book, 1965, pp. 97-98); or how he was the master of the wind: "And unto Solomon (We subdued) the wind in its raging" (21:81).

[2] The Temple was the warranty of the Jews to stay "fixed"; when the Temple was destroyed, they returned to a "nomadic" life (as dispersion) (Guénon, *Le règne*, p. 199).

[3] As Guénon said, the *Hekal* is a "double-square," and so is the base of Poseidon's temple in Atlantis and the Masonic Lodge, with the length (from East to West) double the width (from North to South) (*Symboles fondamentaux*, p. 264). A Masonic tradition describes the "foundation-stone" of Solomon's Temple "as a double cube, every side, except the base on which it stood, being inscribed. The first face of the cube was said to have been engraved by Noah with an instrument of porphyry when the Ark was being built; the second, by Abraham, with the horn of the ram, which was substituted for his son on

Solomon's Temple   163

could be compared to the Hindu subtle channels, *ida* and *pingala*, or to Hermes' two serpents, or to Hercules' columns. The three floors of rooms around the Temple could represent the cosmic "frame," but also the labyrinth[1]; or, from a different point of view, they are a token of the Three Worlds, and the spiral stairs suggest the degrees of universal Existence. The Temple was built on Mount Moriah, an equivalent of the Hindu *Mêru*, *Axis Mundi*; no tool made of iron desecrated the edifice, with bronze being preferred as a reminder of a better Age. The architect was the great master Hiram, or Adoniram, or Hiram-Abif, the Widow's son ("King Solomon sent to Tyre and brought Hiram, whose mother was a widow from the tribe of Naphtali,"[2]). We have here all the symbolic elements of a myth or of an initiatory fairy-tale, and it is easy to understand why Free-Masonry adopted them as a foundation-stone or a corner-stone.

David and Solomon's cycle, like any cycle, would decay too, a future event God already warned Solomon about. The decadence of each cycle described in the *Bible* is characterized by some main elements, which can be found flourishing to an extreme in our modern world: forgetting or even turning away

---

Mount Moriah; the third, with a porphyry tool by Moses; the fourth by Joshua; and the fifth by Hiram Abif, before it was deposited in its final bed at the north-east angle of the Temple" (Alex Horne, *King Solomon's Temple in the Masonic Tradition*, The Aquarian Press, 1972, p. 150).

[1] Deeds, in his essay *The Labyrinth*, was right in saying that "in Egypt and Crete, the Labyrinth represented the ritual center" (*The Labyinth*, p. 34); also, the fact that Virgil called the labyrinth "Troja" implies that the city of Troy was, like Jerusalem, an image of the Center (see W. H. Matthews, *Mazes and Labyrinths*, Dover, 1970, pp. 156-159). It is interesting that till as late as 1844 the labyrinthine figures were called "The Labyrinth of Solomon." With good reason a legend was preserved saying that Aeneas and some Trojan prisoners, whom he had helped to liberate from Greece, came to England and built a city on the banks of the Thames, which he called Caerdroi-Newydd ("New City of Troy"; Caerdroia means in fact the "walls of Troy"), and which was named later London (Matthews 95, 181). Also, the central role of the labyrinth is well known in the case of the Gothic cathedrals (the labyrinth of Chartres is famous).

[2] *3 Kings* 7:13-14.

from God, or replacing Him with idols[1]; promoting what is wrong, unjust, and untrue[2]; corruption and depravation; and the list could continue indefinitely.

David's wrong-doing was just a "personal" episode, which did not affect the Temple, and therefore God's punishment concerned David only[3]; but Solomon's deviations at the end of his life[4] were in direct connection with the Temple, which forced God to warn about the future penalties. The most terrible retribution is, of course, losing the One or Unity, but that is the "normal" destiny of any cycle: at the beginning, during the "Golden Age," Unity is victoriously shining; afterward, due to the fall, division is born and will rapidly grow. Christianity and Islam were confronted with the same division, first a division in two, then into many sects[5]; what appears different in the case of Judaism is that in the Jewish tradition the division is explicitly recorded as God's punishment, and, in fact, the Judaic scriptures were not afraid to describe the entire history as an illustration of God's Breath (retribution and reward) expressed through the doctrine of the cosmic cycles.

"And Jeroboam the son of Nebat, an Ephrathite of Zereda, Solomon's servant, whose mother's name was Zeruah, a widow,

---

[1] These idols in the modern world could be extremely stupid and materialistic or more subtle, aiming at the psyche.

[2] In the modern world to find the "ten righteous men" would be very difficult. There are open appreciations and advertisements for dishonest, untruthful and unjust behaviour as part of the modern "civilization."

[3] *2 Kings* 12:15. Regarding Uriah's wife and David, there is a comparable episode concerning the Prophet Muhammad and Zeinab, the wife of Zeid, his adopted son (*Qur'an* 33:37).

[4] *3 Kings* 11:5-8. However, the Prophet Muhammad strongly declared that Solomon was no idolater: "Solomon was not an unbeliever" (*Qur'an* 2:102). The Islamic tradition says that after Solomon's death the devils, to tarnish his character, fabricated some books of magic and hid them under the throne, and the Jewish people deceived by this devilish forgery believed that Solomon got his power over *djinns*, men and winds through magic; only the Prophet Muhammad restored the truth about Solomon. We see the similarity with the modern ideas about Solomon's hidden "treasure."

[5] "Lo! As for those who make a division in their religion and become schismatics, no concern at all hast thou with them. Their case will go to Allâh" (*Qur'an* 6:159).

Solomon's Temple 165

even he lifted up his hand against the king. And this was the cause that he lifted up his hand against the king: Solomon built Millo, and repaired the breaches of the city of David his father. And the man Jeroboam was a mighty man of valour: and Solomon seeing the young man that he was industrious, he made him ruler over all the charge of the house of Joseph. And it came to pass at that time when Jeroboam went out of Jerusalem, that the prophet Ahijah the Shilonite found him in the way[1]; and he had clad himself with a new garment; and they two were alone in the field: And Ahijah caught the new garment that was on him, and rent it in twelve pieces: And he said to Jeroboam, Take thee ten pieces: for thus saith the Lord, the God of Israel, Behold, I will rend the kingdom out of the hand of Solomon, and will give ten tribes to thee: (But he shall have one tribe for my servant David's sake, and for Jerusalem's sake, the city which I have chosen out of all the tribes of Israel:) Because that they have forsaken me, and have worshipped Ashtoreth the goddess of the Zidonians, Chemosh the god of the Moabites, and Milcom the god of the children of Ammon, and have not walked in my ways, to do that which is right in mine eyes, and to keep my statutes and my judgments, as did David his father. Howbeit I will not take the whole kingdom out of his hand: but I will make him prince all the days of his life for David my servant's sake, whom I chose, because he kept my commandments and my statutes: But I will take the kingdom out of his son's hand, and will give it unto thee, even ten tribes. And unto his son will I give one tribe, that David my servant may have a light always before me in Jerusalem, the city which I have chosen me to put my name there. And I will take thee, and thou shalt reign according to all that thy soul desireth, and shalt be king over Israel."[2]

This episode also reflects a universal theme: Rehoboam, the prince (Solomon's son), and Jeroboam, the servant (the son of

---

[1] We may note that Ahijah is the prophet of the "lost" center, Shiloh, and his intervention alludes to the influence of such centers upon the division of the world.
[2] *3 Kings* 11:26-37.

the widow), with similar names, are the twins of the World's Egg mentioned in various traditions; the division in two of Solomon's kingdom is an image of the Egg's secession into Heaven and Earth, which produces the evolvement of the cycle.[1] A similar event marks the Christian cycle: "Then the soldiers, when they had crucified Jesus, took his garments, and made four parts, to every soldier a part; and also his coat: now the coat was without seam, woven from the top throughout."[2]

The decay continued, in smaller cycles, the Wrath of God continued through its instruments – the foreign armies, namely the Assyrians and the Egyptians: the former will destroy the kingdom of Israel and Samaria[3]; the latter will attack Jerusalem and the Temple.[4] Jeroboam's reign signifies a minor cycle, which well summarizes the greater ones: "Go, tell Jeroboam, Thus saith the Lord God of Israel, Forasmuch as I exalted thee from among the people, and made thee prince over my people Israel, And rent the kingdom away from the house of David, and gave it thee: and yet thou hast not been as my servant David, who kept my commandments, and who followed me with all his heart, to do that only which was right in mine eyes; But hast done evil above all that were before thee: for thou hast gone and made thee other gods, and molten images, to provoke me to anger, and hast cast me behind thy

---

[1] The separation into two kingdoms is marked by the sacrificial killing of Adoniram, king Rehoboam's representative (*3 Kings* 12:18). The whole episode could be connected to the Masonic symbolism.
[2] *John* 19:23.
[3] About the Assyrians as God's instrument (and ignoring their role) see *Isaiah* 10.
[4] *1 Kings* 14:25-26, *2 Chronicles* 12:2, 12:9, 12:5-8. We may ask if the Western Powers that created the modern secular state of Israel were also God's instruments. On the other hand, the instruments will have their turn to be destroyed: "O Assyrian, the rod of mine anger, and the staff in their hand is mine indignation. I will send him against a hypocritical nation and against the people of my wrath will I give him a charge, to take the spoil, and to take the prey, and to tread them down like the mire of the streets … Wherefore it shall come to pass, that when the Lord hath performed his whole work upon mount Sion and on Jerusalem, I will punish the fruit of the stout heart of the king of Assyria, and the glory of his high looks." (*Isaiah* 10:5-6, 10:12).

back: Therefore, behold, I will bring evil upon the house of Jeroboam. ... Moreover the Lord shall raise him up a king over Israel, who shall cut off the house of Jeroboam that day: but what? even now. For the Lord shall smite Israel, as a reed is shaken in the water, and he shall root up Israel out of this good land, which he gave to their fathers, and shall scatter them beyond the river, because they have made their groves, provoking the Lord to anger. And he shall give Israel up because of the sins of Jeroboam, who did sin, and who made Israel to sin."[1]

The division Rehoboam–Jeroboam is also a good example of how the derived (orthodox or heterodox) traditions and centers were established, starting from an orthodox secondary center. In Jeroboam's case,[2] we could see an indication of a counter-initiatory action, related to a former spiritual center (Shiloh).[3] Jeroboam took over Sichem, Abram's spiritual center,[4] and "converted" it into his capital[5]; he sanctioned as well the duality, but even this duality was upside-down, since he established two golden calves (instead of one, as Aaron did),[6]

---

[1] *3 Kings* 14:7-16.
[2] Jeroboam is a usurper. We may recall how in India, Buddhism, a heterodox tradition, facilitated the act of usurpation. After the reign of Kalashoka, the *Shûdra* usurper Mahapadma Nanda takes the power; later, the usurper Chandragupta takes his master king Dhana Nanda's place (Chandragupta's nephew was Ashoka). Guénon said: "Chandragupta was indeed a Shûdra" (*Autorité spirituelle et pouvoir temporal*, Éd. Didier et Richard, 1930, p. 98).
[3] Normally, a religious adaptation is supervised by a spiritual authority (Guénon, *Autorité spirituelle et pouvoir temporal*, p. 54), which was not Jeroboam's case.
[4] "And Abram passed through the land unto the place of Sichem, unto the plain of Moreh. And the Canaanite was then in the land. And the Lord appeared unto Abram, and said, Unto thy seed will I give this land: and there builded he an altar unto the Lord, who appeared unto him" (*Genesis* 12:6-7).
[5] *3 Kings* 12:25.
[6] See Guénon, *Le Roi*, p. 60. "And Jeroboam said in his heart, Now shall the kingdom return to the house of David: If this people go up to do sacrifice in the house of the Lord at Jerusalem, then shall the heart of this people turn again unto their lord, even unto Rehoboam king of Judah, and they shall kill me, and go again to Rehoboam king of Judah. Whereupon the king took counsel, and made two calves of gold, and said unto them, It is too much for

which was an act of dividing the idolatry itself. One golden calf was established in the southern city of Bethel, the famous spiritual center,[1] which was "converted" to the deviated tradition[2]; the other golden calf was established in the northern city of Dan. In this way the two calves parody the two Cherubim and the guardians of the solar gates.[3] No wonder that the Islamic tradition had in its center the proclamation of God's uniqueness, seconded by an indefatigable battle against idolatry (supported with examples from Judaism and the Christian tradition).[4]

Jeroboam's deviation was also an opportunity for God to illustrate how the law of compensation works, because not only does partial disequilibrium belong to the central and overall equilibrium,[5] but any heterodoxy produces as a reaction a divine

---

you to go up to Jerusalem: behold thy gods, O Israel, which brought thee up out of the land of Egypt. And he set the one in Bethel, and the other put he in Dan" (*3 Kings* 12:26-29).

[1] "And Jacob rose up early in the morning, and took the stone that he had put for his pillows, and set it up for a pillar, and poured oil upon the top of it. And he called the name of that place Bethel: but the name of that city was called Luz at the first" (*Genesis* 28:18-19).

[2] This type of "conversion" is common. The Muslims changed Hagia Sophia into a mosque; the Christians changed the Mezquita of Cordoba into a cathedral. On the other hand, Solomon's Temple was completely destroyed.

[3] Today, it is interesting to see the passion of the Western world to display in their cities statues of calves or cows.

[4] "And when Allâh saith: O Jesus, son of Mary! Didst thou say unto mankind: Take me and my mother for two gods beside Allâh? he saith: Be glorified! It was not mine to utter that to which I had no right" (*Qur'an* 5:116); "Praise be to Allâh, Who hath created the heavens and the earth, and hath appointed darkness and light. Yet those who disbelieve ascribe rivals (gods) unto their Lord" (6:1) "Follow that which is inspired in thee from thy Lord; there is no God save Him; and turn away from the idolaters" (6:106).

[5] "As a whole, the equilibrium is composed of the sum of all the unbalanced parts, and each partial disorder concurs, willy-nilly, to a perfect order" (Guénon, *Études sur l'hindouisme*, p. 15). René Guénon stated that any antitraditional, profane and even counter-initiatory actions or forces cannot surpass the individual domain (the "psycho-physical" world) and it is an illusion to think that they can oppose the spiritual order itself. Without their awareness and despite their will, these entities are subjugated to *Spiritus*, the same way everything is, even if unwitting or involuntarily, subjugated to the

intervention in order to preserve the evolvement of the "chain of cycles" and communication with the Center. For this very reason, the prophet Elijah the Tishbite, whose essential function and symbolism are well known, lived in those days in Jeroboam's kingdom[1]; for this reason, Elijah built the symbolic center on Mount Carmel: "Now therefore send, and gather to me all Israel unto Mount Carmel. ... And Elijah took twelve stones, according to the number of the tribes of the sons of Jacob,[2] unto whom the word of the Lord came, saying, Israel shall be thy name: And with the stones he built an altar in the name of the Lord."[3] And his ascent to heaven (including his body) in a chariot of fire[4] could symbolize the definitive departure of the *Shehinah* from the kingdom of Israel and its final destruction.[5]

Divine Will. And they are used, against their will, for the realization of the "divine plan in the human domain." And Guénon added: "If we consider the matter from an overall perspective, and not only in respect to these beings [representing the counter-initiation], we may say that, similar to all the others, they are necessary in their places, as elements of the assembly, and as 'providential' instruments – speaking in a theological language – of the advance of this world through its cycle of manifestation, because in this way each partial disorder, even when it appears as the disorder, concurs necessarily to total order" (*Le règne*, p. 355).

[1] *3 Kings* 17:1. "Then said Elijah unto the people, I, even I only, remain a prophet of the Lord; but Baal's prophets are four hundred and fifty men" (18:22).

[2] We note that the center has nothing to do with the division into two states.

[3] *3 Kings* 18:19, 18:31-32. Even more significant is Elijah's retreat into the cave of Horeb, the Mount of God, where Moses received the tradition. Elijah is a reiteration of Moses for another cycle, and so will be Jesus Christ, since Elijah "went in the strength of that meat forty days and forty nights unto Horeb the mount of God; and Moses "was there with the Lord forty days and forty nights; he did neither eat bread, nor drink water" (*Exodus* 34:28); and Jesus "had fasted forty days and forty nights" (*Matthew* 4:2). The difference is that Moses was "in the top of the mount" (*Exodus* 34:2), while Elijah was in the cave ("And he came thither unto a cave, and lodged there," *3 Kings* 19:9), which, as Guénon explained, illustrates the descent of the cycle.

[4] "Behold, there appeared a chariot of fire, and horses of fire, and parted them both asunder; and Elijah went up by a whirlwind into heaven" (*4 Kings* 2:11).

[5] "Then the king of Assyria came up throughout all the land, and went up to Samaria, and besieged it three years. In the ninth year of Hoshea the king of

The betrayal of the Judaic orthodox tradition, which implied the betrayal of the One and only Principle, motivated the end of Jeroboam's cycle,[1] and illustrates how at the end of times the hordes of Gog and Magog will invade through the fissures of the Grand Wall,[2] an invasion both from inside (the usurpation and the idols) and outside (the Assyrians).

Yet "also Judah kept not the commandments of the Lord their God, but walked in the statutes of Israel which they made,"[3] even though Solomon's Temple supported this cycle to last longer. Rehoboam's cycle is directly related to the Temple as a secondary center (while Jeroboam's cycle was a deviation) and its end is described as a desecration of the center itself.[4]

Assyria took Samaria, and carried Israel away into Assyria" (*4 Kings* 17:5-6). The Assyrian chronicles report that Sargon II (and not his son, Shalmaneser, *4 Kings* 17:3) conquered Samaria in 722 B.C. However, this event ended the kingdom of Israel; the deported Jews were replaced with Asian tribes, and these peoples together with the remainder of the Jews formed the Evangelic Samaritans.

[1] "For they served idols, whereof the Lord had said unto them, Ye shall not do this thing. Yet the Lord testified against Israel, and against Judah, by all the prophets, and by all the seers, saying, Turn ye from your evil ways, and keep my commandments and my statutes, according to all the law which I commanded your fathers, and which I sent to you by my servants the prophets" (*4 Kings* 17:12-13).

[2] *Qur'an* 18:92-99.

[3] *4 Kings* 17:19.

[4] "For the Lord brought Judah low because of Ahaz king of Israel; for he made Judah naked, and transgressed sore against the Lord. And Tilgathpilneser king of Assyria came unto him. ... And in the time of his distress did he trespass yet more against the Lord: this is that king Ahaz. For he sacrificed unto the gods of Damascus, which smote him: and he said, Because the gods of the kings of Syria help them, therefore will I sacrifice to them, that they may help me. But they were the ruin of him, and of all Israel. And Ahaz gathered together the vessels of the house of God, and cut in pieces the vessels of the house of God, and shut up the doors of the house of the Lord, and he made him altars in every corner of Jerusalem" (*2 Chronicles* 28:21-24). "Now in the fourteenth year of king Hezekiah [son of Ahaz] did Sennacherib king of Assyria come up against all the fenced cities of Judah, and took them. ... And Hezekiah gave him all the silver that was found in the house of the Lord, and in the treasures of the king's house. At that time did Hezekiah cut off the gold from the doors of the temple of the Lord, and from

Solomon's Temple

There are, as usually happens, smaller cycles within the greater one, since the evolvement of a cycle is never a linear descent, and there are partial recoveries (even though the overall march of the cycle is a descent).[1] Thus, God through the prophet Isaiah helped Hezekiah, and "Sennacherib king of Assyria departed, and went and returned, and dwelt at Nineveh," and the Temple was saved.[2] By contrast, Hezekiah's son, Manasseh, was God's worst enemy, the inner enemy[3] who attacked the center.[4] And again God sent his menace well in advance: "Therefore thus saith the Lord God of Israel, Behold, I am bringing such evil upon Jerusalem and Judah, that whosoever heareth of it, both his ears shall tingle. And I will stretch over Jerusalem the line of Samaria, and the plummet of the house of Ahab: and I will wipe Jerusalem as a man wipeth a dish, wiping it, and turning it upside down."[5]

In fact, from the beginning of a cycle the menace is present,[6] which means that God's punishment is not instant, but follows the course of the cycle, which is not at all linear but constituted of many cycles, like a mail shirt made of numerous small metal

---

the pillars which Hezekiah king of Judah had overlaid, and gave it to the king of Assyria" (*4 Kings* 18:13-16).

[1] "It must not be forgotten that the cyclic laws apply to different degrees, for periods not of the same extent, which also sometimes encroach on one another; hence the complications which, at first sight, may seem inextricable and which can in effect be resolved only by considering the order of hierarchic subordination of the corresponding traditional centers" (Guénon, *Le Roi*, p. 68).

[2] *4 Kings* 19:5-7, 19:20, 19:35-36.

[3] The Assyrians were the outer enemy. The attack of God and Magog has to be correlated with the interior decay of a tradition.

[4] "And he did that which was evil in the sight of the Lord, after the abominations of the heathen, whom the Lord cast out before the children of Israel. ...And he built altars in the house of the Lord, of which the Lord said, In Jerusalem will I put my name. ... And he set a graven image of the grove that he had made in the house, of which the Lord said to David, and to Solomon his son, In this house, and in Jerusalem, which I have chosen out of all tribes of Israel, will I put my name for ever" (*4 Kings* 21:2, 21:4, 21:7).

[5] *4 Kings* 21:12-13.

[6] Essentially, it corresponds to the Biblical "fall" and gives the "direction" of the cycle's development.

rings. For this reason, Josiah, Manasseh's grandson, "did that which was right in the sight of the Lord, and walked in all the way of David his father, and turned not aside to the right hand or to the left [but stayed in the center]"; Josiah restored the center, since he rebuilt the Temple,[1] marked the center with the Ark,[2] recovered the Lost Word, the Tradition,[3] held the Passover,[4] and tried to reassemble the One and only by extending the regeneration of the Judaic tradition to Samaria[5] and Bethel.[6]

And after that, the end came, because God "turned not from the fierceness of his great wrath, wherewith his anger was kindled against Judah, because of all the provocations that Manasseh had provoked him withal. And the Lord said, I will remove Judah also out of my sight, as I have removed Israel, and will cast off this city Jerusalem which I have chosen, and

---

[1] "And let them give it to the doers of the work which is in the house of the Lord, to repair the breaches of the house, Unto carpenters, and builders, and masons, and to buy timber and hewn stone to repair the house" (*4 Kings* 22:5-6). For this reason, Josiah was part of the Royal Arch Masonry and still is the core of the Irish Royal Arch ritual.

[2] "And said unto the Levites that taught all Israel, which were holy unto the Lord, Put the holy Ark in the house which Solomon the son of David king of Israel did build" (*2 Chronicles* 35:3).

[3] "And Hilkiah the high priest said unto Shaphan the scribe, I have found the book of the law in the house of the Lord. ... And the king sent, and they gathered unto him all the elders of Judah and of Jerusalem. And the king went up into the house of the Lord, and all the men of Judah and all the inhabitants of Jerusalem with him, and the priests, and the prophets, and all the people, both small and great: and he read in their ears all the words of the book of the covenant which was found in the house of the Lord" (*4 Kings* 22:8, 23:1-2).

[4] "Surely there was not holden such a passover from the days of the judges that judged Israel, nor in all the days of the kings of Israel, nor of the kings of Judah; But in the eighteenth year of king Josiah, wherein this passover was holden to the Lord in Jerusalem" (*4 Kings* 23:22-23).

[5] "And all the houses also of the high places that were in the cities of Samaria, which the kings of Israel had made to provoke the Lord to anger, Josiah took away" (*4 Kings* 23:19).

[6] "Moreover the altar that was at Bethel, and the high place which Jeroboam the son of Nebat, who made Israel to sin, had made, both that altar and the high place he brake down, and burned the high place, and stamped it small to powder, and burned the grove" (*4 Kings* 23:15).

Solomon's Temple

the house of which I said, My name shall be there."[1] The overall fate of a cycle cannot be changed by any particular or secondary regeneration and restoration, since only in the supreme Center do the turbulent rotations cease, change becomes immutability and movement rests.

However, the destruction of Solomon's Temple was not the ending end. In 605 B.C., the Babylonian king Nebuchadnezzar captured Jehoiakim, the Jewish king, and many others (including Daniel, the prophet)[2]; in 597 B.C., the second invasion was aimed directly at the Temple, and the center was desecrated (Ezekiel[3] the prophet was then deported)[4]; in 587-6 B.C., the Temple was completely obliterated and the Babylonian captivity started.[5] Yet after 70 years the Jews were allowed to rebuild their center.[6]

---

[1] *4 Kings* 23:26-27.
[2] "And Jehoiakim the king of Judah went out to the king of Babylon, he, and his mother, and his servants, and his princes, and his officers: and the king of Babylon took him in the eighth year of his reign" (*4 Kings* 24:12).
[3] Ezekiel predicted the departure of the *Shekinah*. "Then the glory of the Lord departed from off the threshold of the house, and stood over the Cherubim. And the Cherubim lifted up their wings, and mounted up from the earth in my sight: when they went out, the wheels also were beside them, and every one stood at the door of the east gate of the Lord's house; and the glory of the God of Israel was over them above" (*Ezekiel* 10:18-19).
[4] "And he carried out thence all the treasures of the house of the Lord, and the treasures of the king's house, and cut in pieces all the vessels of gold which Solomon king of Israel had made in the temple of the Lord, as the Lord had said" (*4 Kings* 24:13).
[5] *Jeremiah* 52.
[6] *Jeremiah* 29:10. Again, the instruments of God (the Babylonians) are in their turn punished: "Behold, I will stir up the Medes against them ... And Babylon, the glory of kingdoms, the beauty of the Chaldees' excellency, shall be as when God overthrew Sodom and Gomorrah" (*Isaiah* 13:17-19); see also *Jeremiah* 50:18. We have to keep in mind that the counter-initiatory forces have a similar fate. The Islamic tradition underscores the alternation of punishment and mercy: "So when the time for the first of the two came, We roused against you slaves of Ours of great might who ravaged (your) country, and it was a threat performed. Then we gave you once again your turn against them, and We aided you with wealth and children and made you more in soldiery. (Saying): If ye do good, ye do good for your own souls, and if ye do evil, it is for them (in like manner). So, when the time for the second (of the

If Zerubbabel and Jeshua were the builders of the Temple,[1] Nehemiah the prophet was the builder of the walls of Jerusalem,[2] and we know the importance of the "cutting" in protecting the center.[3] The cycle of Zerubbabel's Temple lasted for half a century, and the signs of the end were, among others, the desecration of the Temple by Antiochus IV Epiphanes, who, in 168 B.C., erected a statue of Zeus in the Holy of Holies,[4] followed by the Maccabean revolt when Jerusalem was liberated and the Temple purified,[5] which ended with Pompey's

---

judgments) came (We roused against you others of Our slaves) to ravage you, and to enter the Temple even as they entered it the first time, and to lay waste all that they conquered with an utter wasting" (*Qur'an* 17:5-7).

[1] "Then stood up Jeshua the son of Jozadak, and his brethren the priests, and Zerubbabel the son of Shealtiel, and his brethren, and builded the altar of the God of Israel, to offer burnt offerings thereon, as it is written in the law of Moses the man of God" (*Ezra* 3:2). "And I will shake all nations, and the desire of all nations shall come: and I will fill this house with glory, saith the Lord of hosts" (*Haggai* 2:7); Haggai was the prophet of the new (second) temple.

[2] "They which builded on the wall, and they that bare burdens, with those that laded, every one with one of his hands wrought in the work, and with the other hand held a weapon" (*Nehemiah* 4:17); the Masonic ritual reflects this statement.

[3] "And they shall build the old wastes, they shall raise up the former desolations, and they shall repair the waste cities, the desolations of many generations" (*Isaiah* 61:4).

[4] We should mention the sayings of Hecataeus of Abdera: "under the rule of nations during later times, namely, of the Persians and Macedonians ... the Jews greatly modified the traditions of their fathers" (see *The Labyrinth*, p. 122); it seems that the Jews of the Dispersion, in the time of Alexander and his successors, accepted some foreign influences (one is considered to be the influence from the cult of Sabazios); certain synagogues contained paintings depicting scenes from the *Old Testament*.

[5] Some thought to detect a Greek influence upon the ritual of *Hanukkah*, which was instituted at that time, since the Jewish law asked the householders to light a lamp at sunset and place it at the door of the house, outside, which was similar to the Greek tradition of placing altars in the streets (see *The Labyrinth*, pp. 161-209); "Moreover king Antiochus wrote to his whole kingdom, that all should be one people, and every one should leave his laws: so all the heathen agreed according to the commandment of the king. Yea, many also of the Israelites consented to his religion, and sacrificed unto idols, and profaned the Sabbath. For the king had sent letters by messengers unto

conquest of Jerusalem and the desecration of the Temple, in 63 B.C. Tacitus described the Temple and its desecration: "There stood a temple of immense wealth. First came the city with its fortifications, then the royal palace, then, within the innermost defenses, the temple itself. ... When the Macedonians became supreme, King Antiochus strove to destroy the national superstition, and to introduce Greek civilization ... Cneius Pompeius was the first of our countrymen to subdue the Jews. Availing himself of the right of conquest, he entered the temple. Thus it became commonly known that the place stood empty with no similitude of gods within, and that the shrine had nothing to reveal. The walls of Jerusalem were destroyed, the temple was left standing."[1]

Just before the end, as normally happens, the Temple was significantly expanded and adorned. Herod commanded this work in 23 B.C. and it continued for three quarters of a century: "In the fifteenth year of his reign, Herod rebuilt the temple, and

---

Jerusalem and the cities of Judah that they should follow the strange laws of the land, and forbid burnt offerings, and sacrifice, and drink offerings, in the temple; and that they should... set up altars, and groves, and chapels of idols, and sacrifice swine's flesh, and unclean beasts ... And whosoever would not do according to the commandment of the king, he said, he should die. ... Now the fifteenth day of the month Casleu, in the hundred forty and fifth year, they set up the abomination of desolation upon the altar, and builded idol altars throughout the cities of Judah on every side; and burnt incense at the doors of their houses, and in the streets" (*1 Macabees* 1:41-55). We witness here the "dispersion," the spread from the Center into multiplicity, from the one and only altar of the Temple of Solomon to innumerable altars in the streets; and it is curious to see how the "festival of lights," which became *Hanukkah*, illustrates the same "dispersion," because initially the light was the one in the Temple, as Josephus said: "So on the five and twentieth day of the month Casleu, which the Macedonians call Apelleus, they lighted the lamps that were on the candlestick" (*Antiquities of the Jews*, book XII, 7:6, *The Life and Works of Flavius Josephus*, The John C. Winston Co., Philadelphia, cca. 1900, tr. William Whiston, p. 367).

[1] *Historiae*, book V. "For Pompey, and those that were about him, went into the temple itself whither it was not lawful for any to enter but the high priest, and saw what was reposited therein, the candlestick with its lamps, and the table, and the pouring vessels, and the censers, all made entirely of gold, as also a great quantity of spices heaped together, with two thousand talents of sacred money" (Flavius Josephus, *War of the Jews*, I, 7).

encompassed a piece of land about it with a wall, which land was twice as large as that before enclosed. The expenses he laid out upon it were vastly large also, and the riches about it were unspeakable."[1] "The temple resembled a citadel, and had its own walls, which were more laboriously constructed than the others. Even the colonnades with which it was surrounded formed an admirable outwork. It contained an inexhaustible spring; there were subterranean excavations in the hill, and tanks and cisterns for holding rain water. The founders of the state had foreseen that frequent wars would result from the singularity of its customs, and so had made every provision against the most protracted siege. After the capture of their city by Pompey, experience and apprehension taught them much. Availing themselves of the sordid policy of the Claudian era to purchase the right of fortification, they raised in time of peace such walls as were suited for war. Their numbers were increased by a vast rabble collected from the overthrow of the other cities."[2]

Tacitus described as well how Titus, in 70 A.D., conquered Jerusalem, a city where a brutal and deadly division reigned among the Jews and the Temple was completely obliterated.[3] Without the center, the Jews were scattered into the world, but some remained in the region and the hope to rebuild the Temple was still alive.[4] At one moment they thought that Hadrian the emperor would help them, but he built instead Aelia Capitolina, and the land was ploughed up. Then, in 132

---

[1] Josephus, *War*, I, 21.
[2] Tacitus, *Historiae*, book V.
[3] The Prophet Muhammad noticed this division in his days, when the Jews of the tribes of Koreidha battled those of al-Aws, al-Nadhîr, and al-Khazraj: "And when we took a covenant from you, shed ye not your kinsman's blood, nor turn your kinsmen out of their homes: then did ye confirm it and were witnesses thereto. Yet ye were those who slay your kinsmen and turn a party out of their homes, and back each other up against them with sin and enmity" (*Qur'an* 2:84-85). Unfortunately, the Christians and the Muslims committed the same "sin."
[4] A Judaic tradition says that the stones of the Temple were not lost but secretly kept by the Holy One, Blessed be He, for the future Temple.

A.D., Shimon ben Kosiba liberated Jerusalem and was seen as Messiah, therefore the Jews called him Bar Kokhba, the "son of the Star"[1]; there are opinions that he rebuilt the Temple,[2] which, in 135 A.D., would be destroyed by Hadrian, who built a Roman temple instead.

The birth of Christianity did not change the status of Jerusalem, which remained the image of the "Center of the World," since where else could the *Avatâra* be born if not in the Center? The Christians continued the Jewish prophets' perspective and considered the Temple's destruction as God's punishment, in concert with the Judaic scriptures and with Christ's sayings[3]: "And Jesus went out, and departed from the temple: and his disciples came to him for to shew him the buildings of the temple. And Jesus said unto them, See ye not all these things? Verily I say unto you, There shall not be left here one stone upon another, that shall not be thrown down."[4] "And when ye shall see Jerusalem compassed with armies, then know that the desolation thereof is nigh. Then let them which are in Judaea flee to the mountains; and let them which are in the midst of it depart out; and let not them that are in the countries enter thereinto. For these be the days of vengeance, that all things which are written may be fulfilled."[5]

The Jews, Christians, and Muslims, all, from a religious viewpoint, disregarded the doctrine of the cosmic cycles, and had in sight the very end and the supreme Center, where everything would be reassembled; although they sometimes waited for an imminent salvation and thought that being Jew, Christian or Muslim was enough to obtain it. Yet the secondary cycles followed the normal path even after the destruction of Solomon's Temple.

---

[1] "I shall see him, but not now: I shall behold him, but not nigh: there shall come a Star out of Jacob" (*Numbers* 24:17).
[2] The Temple is shown on Bar Kokhba's coins.
[3] Jesus referred to Daniel: "When ye therefore shall see the abomination of desolation, spoken of by Daniel the prophet, stand in the holy place" (*Matthew* 24:15).
[4] *Matthew* 24:1-2, *Mark* 13:1-2, *Luke* 21:6.
[5] *Luke* 21:20-22.

A temple as center was built where Helena, Constantine's mother, recovered the True Cross: the Church of the Holy Sepulchre.[1] Desolation reigned over the place where Christ suffered death and was buried,[2] yet, the legend says, Helena and St. Macarius, the Bishop of Jerusalem, cleaned it and discovered the True Cross during the excavations[3]; the church, encompassing the Christian holy sites, included a rotunda called the *Anastasis* ("Resurrection"), which represented a dome over a cylinder.[4] It is said that Helena built another small church, above the Rock, calling it the Church of St. Cyrus and St. John, which later was enlarged and became the Church of the Holy Wisdom (Hagia Sophia), having an octagonal shape; in 614 A.D., the Persians, seemingly, destroyed the church.[5]

---

[1] We should observe that, in the Middle Ages and later, Solomon's Temple was often identified with the Church of the Holy Sepulchre, which became the new "Temple." Let us note that it is no coincidence that Solomon's Temple is said to be either on Mount Sion or on Mount Moriah, and that today nobody knows exactly where the Temple was located.

[2] The site of the Holy Sepulchre was covered with dirt when Hadrian built his Aelia Capitolina in 135 A.D.

[3] The True Cross is, like the Ark of the Covenant and the Holy Grail, a symbol of the Lost Tradition. Likewise Seth, who returned to the Terrestrial Paradise (the Center) and recovered the Holy Grail (retrieving the Primordial Tradition), which allowed him to establish a spiritual center replacing the lost Paradise and being its image, the recovering of the True Cross symbolized the finding of Christ as Center and Tradition, and permitted the foundation of the new temple as center.

[4] It illustrated the Latin or Occidental architectural "style" (see the Pantheon of Rome, already discussed in chapter IV, which in 609 A.D. became the Church of Santa Maria Rotonda). The same "round church" was the New Templar Church in London (consecrated in honour of the Blessed Virgin Mary in 1185 and built over a Roman temple, which has more than one meaning); "And he carved all the walls of the house round about with carved figures of Cherubim and palm trees and open flowers, within and without" (*3 Kings* 6:29) (Bede commented on this: "All the walls of the temple all the way round are all the people of the holy Church laid upon the foundation of Christ; with these he has filled the earth's entire circle," p. 53; we may note that the medieval people did not consider the earth flat, a mischievous invention of the modern world).

[5] There are various modern theories regarding the location of the Temple, and some of them consider the Rock outside the holy site, where the Fort Antonia

Solomon's Temple

There is no doubt that Jerusalem continued to play a "central" role, regardless of who were the players. The Jews thought they could rebuild the Temple and overthrow the Christians when Constantine's nephew, Julian, became emperor. The Christians regarded Julian as Antichrist,[1] since he returned to worshiping the idols; but, there is no surprise here, if we keep in mind how many times the Jews made a similar error. This time, the idols gave the Jews the opportunity to rebuild their Temple. Julian "planned at vast cost to restore the once splendid temple at Jerusalem, which after many mortal combats during the siege by Vespasian and later by Titus, had barely been stormed. He had entrusted the speedy performance of this work to Alypius of Antioch, who had once been vice-prefect of Britain. But, though this Alypius pushed the work on with vigour, aided by the governor of the province, terrifying balls of flame kept bursting forth near the foundations of the temple, and made the place inaccessible to the workmen, some of whom were burned to death; and since in this way the element persistently repelled them, the enterprise halted."[2]

It was God's Will, the Christians considered, and the prophets would have considered it the same thing. Photius, quoting Philostorgius, said: "When Julian bade the city of Jerusalem to be rebuilt in order to refute openly the predictions of our Lord concerning it, he brought about exactly the opposite of what he intended. For his work was checked by many other prodigies from heaven; and especially, during the preparation of the foundations, one of the stones which was placed at the lowest part of the base, suddenly started from its place and opened the door of a certain cave hollowed out in the rock. Owing to its depth, it was difficult to see what was within

---

or the Praetorium of Pilate was; however, many of them agree to locate the Temple where the Rock is. From our point of view these theories have little value, since we consider only the traditional perspective.

[1] Antichrist, it is said, will settle in the Temple of Solomon (Patrice Genty, *L'Antéchrist dans les legendes celtiques*, Études Traditionnelles, 1939, no. 236-237-238).

[2] Ammianus Marcellinus, *Rex Gestae*, XXIII, 1.

this cave; so persons were appointed to investigate the matter, who, being anxious to find out the truth, let down one of their workmen by means of a rope. On being lowered down he found stagnant water reaching up to his knees; and, having gone round the place and felt the walls on every side, he found the cave to be a perfect square. Then, in his return, as he stood near about the middle, he struck his foot against a column which stood rising slightly above the water. As soon as he touched this pillar, he found lying upon it a book wrapped up in a very fine and thin linen cloth; and as soon as he had lifted it up just as he had found it, he gave a signal to his companions to draw him up again. As soon as he regained the light, he showed them the book, which struck them all with astonishment, especially because it appeared so new and fresh, considering the place where it had been found. This book, which appeared such a mighty prodigy in the eyes of both heathens and Jews, as soon as it was opened, showed the following words in large letters: 'In the beginning was the Word, and the Word was with God, and the Word was God.' In fact, the volume contained that entire Gospel which had been declared by the divine tongue of the (beloved) disciple and the Virgin. Moreover, this miracle, together with other signs which were then shown from heaven, most clearly showed that 'the word of the Lord would never go forth void,' which had foretold that the devastation of the Temple should be perpetual. For that book declared Him who had uttered those words long before, to be God and the Creator of the universe; and it was a very clear proof that 'their labour was but lost that built,' seeing that the immutable decree of the Lord had condemned the Temple to eternal desolation."

Photius' description, symbolizing the finding of the "Lost Word," is similar to others belonging to various traditional data, from fairy tales[1] to the *Royal Arch* ritual; it also reminds us of the book of the covenant found in the house of the Lord in the time of Josiah.[2] It suggests the end of the Jewish cycle and the victory of the Christian one, while the Temple's devastation

---

[1] See our *Agarttha, the Invisible Center*, pp. 61-2.
[2] *4 Kings* 22:8, 23:1-2.

should be perpetual.[1] Nonetheless, in 614 A.D., the Persians occupy Jerusalem, and the Jews have hopes that king Chosroes will permit the rebuilding of the Temple[2]; it was a last attempt, since after 15 years the Byzantine emperor Heraclius recaptures Jerusalem.[3]

Heraclius' time represented a new epoch for the Temple, but also for the world. In 610 A.D., Muhammad the Prophet started his mission and in the same year Heraclius became emperor.[4] In 614, the Persians not only occupied Jerusalem but took with them the Holy Cross (which symbolically meant that the Grail was lost)[5]; only in 627 could the Byzantines defeat the

---

[1] This is an exoteric view, and it does not mean that the Temple lost its importance; on the contrary, the climax of the Christian cycle would bring Solomon's Temple in the center. At the same time, the medieval Christian society (a traditional one) highly praised Solomon in initiatory and symbolical songs, stories, and poems, which were not unrelated to the Order of the Temple and Masonry (the Jews and the Muslims had similar stories about him); Solomon's mythical description was comparable to that of Merlin.

[2] They were allowed to return to Jerusalem for a while, but then Chosroes changed his mind.

[3] The Islamic tradition mentions the following foretelling about Heraclius: "The Greeks have been defeated (by the Persians), in the nearer land, and they, after their defeat, will be victorious" (*Qur'an* 30:2-3).

[4] Like he wanted to stress that a new cycle began, Heraclius made Greek the official language of the Byzantine Empire.

[5] The True and Holy Cross was, in a sense, Jerusalem's *palladium*, through which the center identified itself with god, that is, with Jesus Christ. In various traditions, the capture of a holy city was equivalent with the rapture of the city goddess, and Knight gave as examples Lagash (a Sumerian city; a lamentation for Bau, the goddess of Lagash, goes like this: "the august queen from her temple he brought forth./ O lady of my city, desolated, when wilt thou return?"), the devastation of Akkad by the Elamites, when the goddess Nana of Erech was carried off, and, of course, Troy (Odysseus stole the *palladium* of Pallas-Athena); Virgil said that all the gods left their shrines in Troy and that Poseidon was seen by Aeneas destroying his own work (Knight 105, 120, 296) (the taking of Troy, the fight and the punishment of its citizens, should be compared to the taking of Jerusalem by various armies; in fact, the conquest of a holy city reflects the same symbolism). We remember the destruction of Shiloh, when the Philistines abducted the Ark of the Covenant; also, we should mention the great fire of 1194 that destroyed the Romanesque basilica of Chartres; "when the people of Chartres were unable, for a few days, to see the most holy relic of the Blessed Virgin, this was experienced by them

Persians at Nineveh and one year later (when Muhammad captured Mecca) Heraclius brought back the relics of the True Cross; in 629, he recaptured Jerusalem and took the title of Basileus.[1]

Heraclius saw himself as the emperor preparing the Second Coming and Heavenly Jerusalem and disregarded completely the laws of the cosmic cycles[2]; yet, as normally happens, a few years before his death the Christian reign ended in Jerusalem.[3]

---

as an incredible pain and sadness" (Burckhardt, *Chartres*, p. 83). The basilica of Chartres was the center of the cult of Mary in France, if not in Western Europe, and the fire and disappearance of the relics (the main relic was the shirt that the Virgin Mary was said to have worn at the birth of Christ) were considered as a sign of divine wrath, when the Virgin had abandoned her shrine (similar to the *Shekinah* leaving the Temple); the miraculous recovery of the relic was therefore interpreted as a sign from the Virgin that "she wanted a new and more beautiful church to be built in her honour" (the recovery of the relic literary meant the rebirth of the city) (Simson 159-164).

[1] Heraclius was viewed as the first Crusader and the restorer of Christianity (Chosroes was depicted as the Antichrist); he was considered to be "refounding of Constantine's Christian empire." The symbolism of his deeds is obvious: the defeat of the God and Magog, the recovering of the Tradition (the Holy Cross), the retrieving of the center (Jerusalem) and the re-establishing of the traditional order. It is interesting to note that in a contemporary poem the Cross was compared to the Ark of the Covenant, but considered more powerful. The restitution of the Cross to the Church of the Holy Sepulchre occurred on March 31, 630, and Heraclius' entry into Jerusalem imitated Christ's entry on Palm Sunday and was viewed as the beginning of a new era; Heraclius was the first Byzantine emperor to enter Jerusalem.

[2] The Christians believed Christ was coming soon (Kemp 6); in fact, Jews and Christians were continuously waiting for the Last Judgement, which could occur at any moment, and so they could not take any interest in the doctrine of the cosmic cycles. However, there are in the Judaic tradition sufficient allusions to the doctrine of the cosmic cycles; see *Isaiah* for example, or *Amos*: "For, lo, I will command, and I will sift the house of Israel among all nations, like as corn is sifted in a sieve, yet shall not the least grain fall upon the earth ... In that day will I raise up the tabernacle of David that is fallen, and close up the breaches thereof; and I will raise up his ruins, and I will build it as in the days of old" (*Amos* 9:9, 9:11).

[3] Luc Benoist presented the same idea when he said: "the historians' repugnancy to take in consideration the cycle comes from Christianity"; "Christianity transforms an esoteric truth in a historic fact ... it gives to the relativity of history a characteristic of absoluteness ... it creates at the same

Solomon's Temple

Nonetheless, we should recall that Heraclius played an esoteric and mysterious role as well in the Christian tradition as in the Islamic one. Beside his "heretical" views[1] and the alleged attraction to Islam,[2] Heraclius was perceived as a guardian of Tradition. It is said that Abû Bakr, shortly after he became Caliph, sent (in 632 A.D.) two ambassadors to the Byzantine emperor and they had confidential meetings; during such a meeting (in the middle of the night) Heraclius showed to the ambassadors a "treasure"[3]: a chest containing pieces of silk, each having a portrait painted on it, starting with Adam, Noah and Abraham and continuing with Moses, Aaron, Lot, Isaac, Jacob, Ishmael, Joseph, David, Solomon and Jesus; in the end, the emperor showed them a figure that the ambassadors recognized to be Muhammad the Prophet.[4]

---

time the myth of indefinite progress" (*Le retour aux cycles*, Études Traditionnelles, 1970, no. 421-422). Indeed, it is well known how St. Augustine refuted the theory of the cosmic cycles, since "Christ died once for all for our sins; and in rising from the dead he is never to die again ...The ungodly will walk in circle; not because their life is going to come round again in the course of those revolutions which they believe in, but because the way of their error, the way of false doctrine, goes round in circles" (*City of God*, bk. XII, ch. 14); yet we have to do justice to St. Augustine, since he criticized a theory which believed in an "eternal return," a "repetition of history." See also Kemp.

[1] He promoted "monothelitism," seemingly trying to surpass the "historicism" side.

[2] It is said that Heraclius responded favourably to Muhammad's invitation to embrace Islam.

[3] Heraclius' "treasure" was, it is said, only a copy of the famous treasure of Adam, which means that this "imperial" tradition is a secondary one.

[4] For Vâlsan, this "treasure" could be an Occidental depot of Atlantean origin; Vâlsan compared Heraclius' chest to the Ark of the Covenant and he also referred to the ark (*tâbût*) in which Moses as a new born was hidden (this last reference is very rich in symbolic meanings, since it alludes to the birth of an *avatâra*) (see Michel Vâlsan, Études Traditionnelles, 1962, no. 371, 374, 1963, no. 375). "Then said to them their prophet [Samuel], The sign of his kingdom is that there shall come to you the ark with the Shekinah in it from your Lord, and the relics of what the family of Moses and the family of Aaron left; the angels shall bear it. In that is surely a sign to you if ye believe" (*Qur'an* 2:247); this ark, the traditional Islamic commentators have said, contained the images

In a way, the last portrait warned Heraclius about the near future of Jerusalem: in 638 A.D., Omar captured Jerusalem and allowed the Jews to return; however, the Muslims showed the same unawareness with respect to the doctrine of the cosmic cycles and promoted the theory of an uninterrupted Islamic progress.[1] Between 687 and 691 A.D., the Caliph Abd al-Malik built the Dome of the Rock (*Masjid Qubbat As-Sakhrah*), which is not a mosque and replaced Solomon's Temple, the same way the Church of the Holy Sepulchre did.[2]

---

of the prophets and was sent down from heaven to Adam, and eventually came to the Israelites.

[1] Since the goal of a religion is Paradise, that is, the Center, and the salvation of the soul, this struggle against sins and devils, as well as the conversion of the nations, is seen as progress. This mainly spiritual "progress" (in fact a return to the primordial order) was misunderstood by common people, by kings and priests, and later was aped by the profane world and occultism. Today the idea of "progress" is so deeply inoculated into the mentality of the world, that even when there are obvious cyclic disasters, it is said: "we are making progress." However, the *Qur'an* mentions more about "compensation" than progress: "O ye who believe! Whoso of you becometh a renegade from his religion, (know that in his stead) Allâh will bring a people whom He loveth and who love Him" (5:54) (this was considered a foretelling, since after Muhammad's death, a significant number of Arabs returned to their idols or to Judaism or Christian religion, while others became Muslims).

[2] "By the sixth century various symbols, relics and even holy sites once associated with the Temple had migrated to the Holy Sepulchre complex, and early representations on coins and pilgrims' souvenirs represent the front elevation of the Temple (or the Holy of Holies) and the church of the Holy Sepulchre in a similar schematized manner, despite their very different ground plans." (Bede L).

VII

THE ROCK

Most assuredly, the rock or stone was, as we have already mentioned, an important symbol of the center. René Guénon in his *Le Roi du Monde* devoted an entire chapter to it,[1] stressing that the name *Beith-El* ("house of God") was applied not only to the place, but to the stone itself, and when we talk about the "cult of stones," which was common to many ancient people, we must understand this cult not addressing the stones, but the Divinity for whom they were residence. Such special stones illustrated the center manifested as temple, even though, in time, they became "idols," like the stones of the pre-Islamic Arabs.[2]

---

[1] *L'omphalos et les bétyles.*
[2] For example: Manat was a large stone worshiped in the territory of the Hudhail tribe (between Mecca and Medina), demolished by Saad in the eighth year of Hegira; Allât was a rectangular stone and the "idol" of the tribe of Thakîf, having a temple in Nakhlah (the idol was demolished by Muhammad's unbending order; its loss was perceived by the tribe in a similar way as the Trojans perceived the loss of their *palladium*); in some cases, the divinity was identified with a particular part of a natural rock. There were other "idols," mentioned in the *Qur'an*, of antediluvian origin, worshiped under the form of a man (Wadd), a woman (Suwâ), a lion (Yaghût), a horse (Yaûk), and an eagle (Nasr); we see the similarity with the Mesopotamian gods, with Ezekiel's tetramorph, and with the symbols of the four Christian Evangelists, which shows how people's mentality, in accord with the situation of the cycle's evolvement and the spiritual influences' presence or absence, makes a stone or a rock become an "idol." Gaudefroy-Demombynes mentioned that in the pre-Islamic pilgrimage each station (*wuqûf*) was marked by a stone or a mountain (*Le pèlerinage à la Mekke*, Librairie Orientaliste Paul Geuthner, 1923, p. IV).

The rock is present in the Judaic and Christian traditions, alluding to the symbolism of the center: "He shall cry unto me, Thou art my father, my God, and the rock of my salvation"[1]; "I will liken him unto a wise man, which built his house upon a rock"[2]; "And I say also unto thee, That thou art Peter, and upon this rock I will build my church."[3] It is only normal to see the symbolism of the rock maintained in the Islamic tradition and the building of the Dome of the Rock should be perceived as a profound act of spiritual recognition of the rock as a center.[4] All the traditional data regarding the rock in the center of the Dome sustain this reality: this is the rock where Abraham was willing to sacrifice his son Isaac; this is the rock where Jacob dreamed about the heavenly ladder; this is the "foundation-stone" upon which the Ark of the Covenant was placed in the Holy of Holies[5]; this is the place from which Prophet Muhammad ascended to Heaven[6]; this Rock under the Dome was the first praying direction for Muslims before Mecca.[7] What is impossible to understand from an exoteric point of view, appears, from the esoteric perspective, as a shining and immutable truth: the human factor imposed various "garments" upon the Rock, as it imposed upon any other images of the center, while, in reality, the essence of it is one and only, reflecting the one and only Tradition and the one and

---

[1] *Psalms* 89:26.
[2] *Matthew* 7:24.
[3] *Matthew* 16:18.
[4] The symbolism of the Rock as center is stressed also by the cave that exists under it, connected to a well; as Burckhardt said, "the cave under the rock is like the heart or innermost conscience of man" (*Art of Islam*, p. 10). The symbolism of the rock is present in the *Qur'an*: "And when Moses asked for water for his people, We said: Smite with thy staff the rock. And there gushed out therefrom twelve springs (so that) each tribe knew their drinking-place" (2:60).
[5] The "foundation-stone," *Even ha-Shethiyah*, is "the rock from which the world was woven."
[6] Burckhardt, *Art of Islam*, p. 10.
[7] In the Judaic tradition, the *Mishnah* affirms that the prayer should be made by directing the heart towards the Holy of Holies; sometimes it is said that this Holy of Holies belongs to the Heavenly Temple.

only Principle; therefore a mosque could be transformed in a church, or a church into a mosque; therefore the same rock could be veiled with "myths" belonging to different traditions, serving the same invisible purpose.[1]

From the same esoteric perspective, the Islamic tradition, as the last revealed one, used the best terminology allowing us to express the universal and *principial* truth beyond any distinction and specific form. For this reason, René Guénon, when he described the Unity, could affirm: "This luminous spherical form, indefinite and not closed, with its alternations of concentration and expansion (successive from the viewpoint of manifestation, but in reality simultaneous in the 'eternal present') is, in the Islamic esotericism, the form of the *Rûh muhammadiyah*; this is the total form of 'Universal Man' that God commanded the angels to adore."[2] For the same reason, the appellations "Islam" and "Muslim" have a universal essence, as any traditional man realizes their meaning, regardless of the traditional form he belongs to; "Related to this, we should recall that the proper meaning of the word *Islâm* is 'submission to the divine Will'; hence it is said, in certain esoteric teachings, that every being is *muslim*, in the sense that there is clearly none who can elude that Will, and accordingly each necessarily occupies the place allotted to him in the Universe as a whole,"[3] and again, for the same reason, when the great seer Ibn 'Arabî declared that "the Christians and, generally speaking, all 'the men of the scriptures' do not change their religion when they

---

[1] It was admitted that the Rock is a token of the close relation between Jewish, Christian and Muslim traditions, which does not mean that there are not continuous futile polemics trying to prove that the Rock was primarily a Christian symbol, or a Jewish one, or an Islamic one.

[2] René Guénon, *Le symbolisme de la croix*, p. 44. In accord with Guénon's sayings, Michel Vâlsan wrote that in the Supreme Center of the Primordial and Universal Tradition reigns the primordial Muhammadian Being, who corresponds to primordial Manu and to *Melki-Tsedeq* (*L'Islam et la fonction de René Guénon*, Les Editions de l'Oeuvre, 1984, p. 178). Ibn Arabî calls the Supreme Center "the Sublime Assembly" and the Islamic community is its external form, similar to the Judaic tradition where Knesseth-Israel here on earth is the expression of the celestial Knesseth-Israel.

[3] Guénon, *Le symbolisme de la croix*, p. 135.

become Muslims," he referred to the doctrine of Unity, and in this sense and only in this sense we should understand Charles-André Gilis' expression, "the universal spirit of Islam."[1]

"The doctrine of Unity (*Et-Tawhîd*)," René Guénon wrote, "that is, the affirmation that the Principle of all existence is essentially One, is a fundamental point common to all orthodox traditions," while "only in descending toward multiplicity differences of form appear, the modes of expression themselves then being as numerous as that to which they refer, and susceptible to indefinite variation in adapting themselves to the circumstances of time and place."[2] The Islamic tradition, as the last orthodox tradition descended on earth before the end of times, affirms most openly and clearly that "the doctrine of Unity is unique,"[3] that is, this doctrine is everywhere and always the same, unchangeable like the Principle, independent of any multiplicity and of all the changes that influence the contingent applications.[4]

---

[1] Charles-André Gilis, *L'Esprit universel de l'Islam*, Al-Bouraq, 1998, p. 205. In the same way, the Jewish prophets envisaged the Temple at the end of times as a Temple for all nations; see, for example: "And it shall come to pass in the last days, that the mountain of the Lord's house shall be established in the top of the mountains, and shall be exalted above the hills; and all nations shall flow unto it" (*Isaiah* 2:2-4).

[2] René Guénon, *Aperçus sur l'ésotérisme islamique et le Taoïsme*, p. 37.

[3] *Et-Tawhîdu wâhidun*. As Guénon said, in Islam, the statement of Unity is expressed in the most explicit way and so adamant that it seems to absorb all the other statements. "Moreover, this tendency increases with the advance in the development of a cycle of manifestation because this development is itself a descent into multiplicity, and because of the spiritual obscuration that inevitably accompanies it. That is why the most recent traditional forms are those which must express the affirmation of Unity in a manner most visible to the outside; and in fact this affirmation is nowhere expressed so explicitly and with such insistence as in Islam, where, one might say, it even seems to absorb into itself all other affirmations" (*Aperçus sur l'ésotérisme islamique*, p. 39).

[4] Guénon, *Aperçus sur l'ésotérisme islamique*, p. 38. René Guénon was, in his whole work, a "servant" of this Unity (his Islamic name is *Abdel Wahed*, the "servant of the Unique"); he always tried to stress the common origin of the various traditional forms, instead of pointing out the apparent differences, as many are doing today ("what generates division must be banished and what unites must be preserved," *Franc-Maçonnerie*, II, p. 299). His illustrious predecessor, the greatest spiritual master Ibn 'Arabî, did the same thing:

With the decay of the cycle and the increasing distance that separates the world from the Principle (the distance from center to circumference), this truth is forgotten, mainly because human beings live in an extreme multiplicity, and therefore the most recent traditional forms have the duty to affirm Unity as explicitly as possible.[1] Moreover, even if we consider not various but one tradition, such as the Islamic one for example, we will find the human factor and the historic circumstances striving to conceal the essential reality.[2] Apparently, the Caliph Abd al-Malik built the Dome of the Rock trying to attract the pilgrims from Mecca to Jerusalem during the conflict with his adversary Abdullah ibn al-Zubayr, and Jerusalem was closer to Damascus than Mecca[3]; he also wanted to challenge the Church of the Holy Sepulchre.[4] In fact, the essential reason was to

---

"rather than focusing on the external differences or apparent contradictions among various hadîth ... Ibn 'Arabî typically – one might say 'ecumenically' – concentrates on conveying the spiritual meaning and intentions implicit in each Prophetic saying, pointing to a level of understanding unifying what might otherwise be seen as differing or conflicting expressions. (This approach mirrors his more general attitude to the various Islamic sects and schools of law, and ultimately to the observable diversity of human religions and beliefs)" (*The Meccan Revelations*, I, p. 315, note of James W. Morris); Ibn 'Arabî's perspective also illustrates his profound understanding of the Universal Man, as integrating the non-manifestation and the manifestation, the divine and the human, with their characteristics.

[1] Of course, at the beginning of the present *Manvantara*, there was no need to express the affirmation of Unity. On the other hand, today, the modern man, consumed by the reign of quantity, understands almost nothing of the doctrine of Unity; and even if he accepts the existence of three "monotheist" religions, he cannot understand that it is about the one and the same Principle, beyond any duality, or that other traditions, like the Hindu or the Chinese one, are not "polytheist."

[2] As Guénon said, the "historical method" was specifically invented to impede any attempts to find the real causes of the various historical events. And he is right when saying that the history presented to the public at large is not the true history (and we could give thousands of examples from our present times) (Guénon, *Articles et comptes rendus*, p. 112, and *L'Erreur*, p. 404, about the "falsification of history," as part of a designed plan).

[3] Gaudefroy-Demombynes 27, Burckhardt, *Art of Islam*, p. 10.

[4] The Prophet has warned about such rivalry: "And as for those who chose a place of worship out of opposition and disbelief, and in order to cause dissent

stress the symbolism of the unique center and to acknowledge that *Al-Haram El-Sharif* was indeed a "holy land" sheltering the center[1]; regardless of any conjectural element, the Dome of the Rock was erected fundamentally to mark the Center and the House of God.[2]

On the other hand, the development of the Islamic civilization was not a linear progress, but, as any other civilization, it suffered human influence and obeyed the law of the cosmic cycles. Therefore, we should not be surprised to see that three of the first four Caliphs died by assassination[3] or that so many secondary cycles evolved and different Muslim factions

---

among the believers, and as an outpost for those who warred against Allâh and His messenger aforetime, they will surely swear: We purposed naught save good. Allâh beareth witness that they verily are liars" (*Qur'an* 9:107-108).

[1] Apparently, the fact that St.-Denis became the spiritual center of France (which today is a "town of desolation"), strongly connected to the royal power of the Capetians, and a very important pilgrimage site, was due to Abbot Suger's ambition to surpass the other holy sites of the West (like Compostela) (Simson 64, 81, 113); but the influence of St. Bernard on Suger suggests that, in fact, there is no question of rivalry or ambition, but that St.-Denis was indeed a representation of the center.

[2] Since the Arabs, like the Jews, were nomads, the building of permanent sanctuaries required foreign craftsmen (Muslims or not), like in the case of Solomon's Temple. Burckhardt considered that "the interior of the sanctuary [the Dome of the Rock] feels more Byzantine or Roman than the exterior" and "it is possible, and even likely, that the plan based on a star-shaped polygon is a Byzantine legacy which, in its turn, has a Platonic and Pythagorean antecedent in antiquity" (*Art of Islam*, p. 12). For Burckhardt, "there is no doubt that the builders of the Dome of the Rock saw in it an image of the spiritual center of the world; granted that this center is symbolized, for Muslims, by the Kaaba, nevertheless Jerusalem, and Mount Moriah in particular, has always been considered as an avatar of this same center" (*Art of Islam*, p. 12).

[3] Omar was assassinated in 644 by a Persian soldier; Othman was killed in 656 by rebels (during his reign many provinces revolted against his rule: Armenia, Georgia, Iran, Afghanistan, Pakistan, Egypt); in 661, Ali was murdered by a Kharijite. As Ibn 'Arabî said, the veritable Caliph is only the one who God chooses to govern His servants, and not the one elected by the people (Ibn 'Arabî, *L'Alchimie*, p. 45), which makes all the difference and accounts for the "human" behaviour.

have competed with each other.¹ It would be, undoubtedly, a mistake to confuse the human with the divine, to believe that a traditional civilization such as the Islamic one, which was founded at the end of the *Kali-yuga*, could represent a paradisiacal society, or, on the contrary, to think that all these inherent imperfections could in any way touch or affect the spiritual doctrine and the Islamic sacred essence.² For Frithjof Schuon, "Islam emerged under the form of an epopee: or a heroic history written with the sword. ... The Arab – and the man made an Arab by Islam – has four poles, that is, the desert, the sword, the woman, the religion"; and Schuon recognized the "human factor" when he said that "Islam possesses essentially a political dimension that was not part of Christianity," and that politics by its nature creates division. He adds: "the drama of the Companions [of the Prophet] is that of human subjectivity."³

---

¹ The most obvious cycles are related to the Omayyad Caliphate of Damascus and Abbasid Caliphate of Baghdad. Abdullah ibn al-Zubayr, who was Aisha's son, did not accept the second Omayyad caliph, Yazid, and proclaimed himself the righteous caliph, ruling over Iraq, southern Arabia, Syria, and parts of Egypt, though in the end he was left by Marwan, the fourth Omayyad caliph, with only the Hejaz (*al-Hejaz*, "the barrier," contained Mecca and Medina). To illustrate the human aspect of Islam, we should add that the Shi'a Muslims considered Yazid the worst tyrant who murdered Husayn ibn 'Ali, but the Sunni Muslims also agreed that he did not follow the Islamic principles and committed a lot of crimes. Husayn ibn 'Ali was Hasan ibn 'Ali's brother, both being the grandsons of the Prophet Muhammad (Hasan was considered the Fifth righteous caliph following Ali; his wife poisoned him, instigated by Mu'awiyah, the first Omayyad caliph, who wanted to assure the succession to his son, Yazid).

² "And when thy Lord said unto the angels: Lo! I am about to place a viceroy in the earth, they said: Wilt thou place therein one who will do harm therein and will shed blood, while we, we hymn Thy praise and sanctify Thee? He said: Surely I know that which ye know not" (*Qur'an* 2:30).

³ *Images d'Islam*, Études Traditionnelles, no. 432-433, 1972. Schuon has acknowledged the fight between the Bedouins and the Qurayshits at the birth of Islam, and reviewed other "violent" events. However, in the same article, he invented two expressions which are not only barbarisms but essentially wrong: "metaphysico-mystic" and "eso-exoterism" (even their form is typically profane and dear to the modern world); Marco Pallis and Jean Borella enthusiastically supported the second expression, albeit René Guénon clearly

There are critics who point out that Jerusalem is never mentioned by name in the *Qur'an* and the "Farthest Mosque" could be any place[1]; indeed, Jerusalem is suggested just symbolically in connection to the Nocturnal Journey: "Glory be to Him who made His servant go by night from the Sacred Temple to the Farthest Temple whose surroundings We have blessed."[2] Yet knowing the circumstances in which the Islamic tradition emerged and especially the fact that it is the "seal" of the previous traditions and the last tradition revealed, there should be no doubt that the Farthest Mosque represents Jerusalem.[3] The Farthest Mosque as center could be compared

---

explained (and Michel Vâlsan later) how the Islamic tradition marks more visibly than any other traditions the distinction between esotericism and exotericism (*Aperçus sur l'ésotérisme islamique et le Taoïsme*, Gallimard, 1973, pp. 13, 21). However, we have to make allowance for Schuon, since he, in a way, only exaggerated René Guénon's sayings, though this exaggeration crossed the line: Guénon, discussing the difference of mentality and nature between Arabs and Persians, underlined the "universality" of Islam and that Shi'ism is not exclusively Persian, but, in a sense, all the Muslims are more or less Shi'a; with respect to Sufism (the Muslim esotericism), this applies as well to Arabs as to Persians, since the Prophet Muhammad taught the "secret science" both to Abu-Bekr and Ali, the Arabic schools being attached to the former and the Persian schools to the latter. The main difference between the two schools, Guénon explained, is that the first ones are more pure intellectually and metaphysically, while the second ones present an esotericism dressed in a more "mystical" form (in the Occidental sense of the word) (*Études sur l'hindouisme*, pp. 120-121).

[1] The *Bible* contains hundreds of references to Jerusalem, yet the word "Jerusalem" appears nowhere in the *Qur'an*. On the other hand, the *hadîth* contain references to *Bayt al-Maqdis* ("House of the Holiness"), that is, Jerusalem.

[2] *Qur'an* 17:1.

[3] The Prophet Muhammad had good reasons to consider Jerusalem the direction of prayer (*Qibla*), which remained so for sixteen and a half months after the Nocturnal Journey and his ascendance to Heaven (*al- Mi'râj*, "ladder") (the opinion that Muhammad chose Jerusalem to attract the Jews of Yathrib to his side has, of course, nothing to do with the real motive). By the time the Caliph Omar captured Jerusalem from the Byzantines in 638 the direction of prayer was already toward Mecca; when Omar first visited the ruined Temple Mount, he prayed south of the ruins of the Temple, toward Mecca. However, the symbolism of the center remained unchanged. It is

to the "Great Extreme" (*Tai Ji*) of the Far-Eastern tradition, which contains Heaven (*Tian*) and Earth (*Di*),[1] while the two "temples," the Sacred and the Farthest, indicate the one and only Center, in the same way that Shambhala (situated at the North Pole and being the "farthest") and Agarttha illustrate the same Center. Some Muslims, at the beginning of Islam, understood the Farthest (*Al-Aqsa*) Mosque as a symbolic place or even a heavenly one, but it is not difficult to admit that, in fact, this was nothing else than an effort to comprehend the complex significance of the center[2]; for these traditions the Journey was from the *Masjid al-Hâram* (the Kaaba) to the "furthest point" (*al-darâb*) identified to the "Inhabited House" (*al-Bayt al-Ma'mûr*), the heavenly Temple of Abraham[3] situated, as Ibn 'Arabî stated, in the seventh heaven.

In 687, the fifth Omayyad caliph, Abd al-Malik ibn Marwan, whose capital was in Damascus, started to build the Dome of the Rock, and his son, Caliph al-Walid, built the *Al-Aqsa* Mosque (*Al-Masjid El-Aqsa*) between 705 and 715, at the southern part of *Haram el-Sharif* (the Noble Sanctuary), making thus explicit the Farthest Temple.[4] No doubt, Mecca (with the Kaaba) represents the spiritual secondary center for the Islamic tradition, yet Jerusalem, which is essential for the Judaic and Christian traditions, remained necessarily a crucial center for the Prophet Muhammad, since the new and last revealed tradition was chosen to integrate both the pre-Islamic elements (Mecca) and the post-Abrahamic ones (Jerusalem)[5]; and the Prophet

---

interesting that "according to a tradition, Jerusalem is called 'the farthest' because it is exactly the center of the world" (Wensinck, *The Navel*, p. 22).

[1] The Great Extreme is identical to the Great Unity (*Tai Yi*). As the Being is the affirmed Non-Being, so *Tai Ji* is the affirmed *Wu Ji*.

[2] We may note that, even though Jerusalem was not a capital-city for the Muslims, as it was for the Jews and the Christian Crusaders (between 1099 and 1187), or as Damascus, Baghdad and Constantinople became, it was in Jerusalem where Mu'awiyah, the first caliph of the Omayyads, was elected.

[3] *The Meccan Revelations*, I, pp. 322, 347.

[4] The Mosque is the second oldest one after the Kaaba.

[5] As Gaudefroy-Demombynes said, "[Muhammad] wanted to join the [pre-Islamic] rites to a religion which would renew and unite the Judaism and the

made clear this "double center" when he completed the nocturnal journey (*isrâ*) from Mecca to Jerusalem.¹

We described earlier how the spiritual influences come down with the celestial Ray and are propagated from the absolute Center into the center of each world and subsequently into the secondary spiritual centers following an uninterrupted flow, which is also the Hesychastic initiatory way, where the uninterrupted Prayer of the Heart compels the intelligence, that is, the spiritual influence, to descend into the heart: "Lord [the supreme Center], Jesus Christ [the celestial Ray], Son of God [the center of the world], have mercy [the spiritual influence] on me, the sinner [human being]."² The Nocturnal Journey (*al-Isrâ*) is an archetypal and universal initiatory journey following the same path, but, of course, in an ascending way, having two symbolical stages: the Lesser Mysteries (*Isrâ'*, from Mecca, the Islamic center, to Jerusalem, the Terrestrial Paradise) and the Greater Mysteries (*Mi'râj*, from Jerusalem to Heaven).³ There

---

Christian religion and which would be the true religion of Abraham, the ancestor of the Arabs" (p. V).

¹ In the Christian tradition, such a "double center" is both Jerusalem-Rome and Jerusalem-Constantinople; in this case too Jerusalem was the Farthest Temple and similar to the Prophet's journey from Mecca to Jerusalem the Crusaders and the pilgrims headed to Jerusalem.

² The Prayer of the Heart follows Solomon's saying: "I sleep, but my heart waketh" (*Song of Solomon* 5:2).

³ We notice the metaphysical and symbolical importance of the "night." René Guénon, in his essential article *Les deux nuits* (*Initiation et réalisation spirituelle*, p. 239 ff.), presented the two extremes of the hierarchy regarding the symbolic meaning of the night, that is, the night as inferior and superior darkness (tenebrae): the superior darkness symbolizes the non-manifestation (the *principial* state of the non-manifested possibilities); the inferior darkness symbolizes the chaos (the pure potentiality). For the human being, *Corpus* is the inferior night and *Spiritus* the superior night. Guénon said: "In the Islamic tradition, the two 'nights' are represented respectively by *laylatul-qadr* and *laylatul-mirâj*, corresponding to a double movement 'descendant' and 'ascendant': the second is the nocturnal ascension of the Prophet, namely the return to the Principle crossing the various 'heavens' which are the superior states of the being, while the first is the night when occurs the descent of the *Qorân* and this 'night,' following Mohyiddin Ibn 'Arabî's commentary, is the body of the Prophet." Guénon also underlined this particularity that the *Qur'an* descended not in the mind but in the body, similar to the Christian

have been vain discussions trying to determine if the journey was a physical one, or was a dream, or a symbolic teaching; in fact, the Prophet's Journey was, as we said, an archetypal and universal spiritual realization, and due to its universality, 600 years later, Dante could describe in Christian clothes a very similar voyage. The main difference is, of course, that Dante, even if he commences his journey also from Jerusalem,[1] is, in fact, lost in the tenebrous forest (*selva selvaggia*), representing a *tamasic* center,[2] while the Prophet Muhammad is in Mecca, a spiritual center, and since he is the messenger of God, his voyage is an archetypal one,[3] a model to be followed by a neophyte like Dante.[4]

Dante's initiatory voyage is similar to the *Kundalinî*'s itinerary, from *chakra* to *chakra*, and follows in detail the paradigmatic

---

saying: *Et Verbum caro factum est* ("the Word was made flesh," *John* 1:14), which stresses the "human" aspect (indeed, the two "nights" are always present together).
[1] "The central point where he [Dante] placed himself corresponded geographically to Jerusalem (Guénon, *L'ésotérisme de Dante*, p. 64). See also Miguel Asín Palacios, *L'Eschatologie musulmane dans la Divine Comédie*, Archè, 1992, p. 20.
[2] "Nel mezzo del cammin di nostra vita,/ Mi ritrovai per una selva oscura,/ Chè la diritta via era smarrita./ Ah! quanto, a dir qual era, è cosa dura,/ Questa selva selvaggia e aspra e forte,/ Che, nel pensier, rinnova la paura!"
[3] For Ibn 'Arabî, the Nocturnal Journey is "above all an archetypal symbol of the highest, culminating stages in the inner, spiritual journey" (*The Meccan Revelations*, I, p. 201); Ibn 'Arabî described more than once his own voyage, as an inner spiritual realization; "my voyage was only in myself."
[4] At the beginning of the voyage both the Prophet Muhammad and Dante are asleep. Dante affirms: "Io non so ben ridir com'io v'entrai,/ Tant'era pieno di sonno a quel punto,/ Che la verace via abbandonai" (*Inf.* I, 10); this sleep, characterizing the "outside darkness" in which Dante drowned, is equivalent to the Platonian amnesia; it is the sleep of the Self, opposite to the Hesychastic spiritual "awakness," to the state of "profound sleep" of the Hindu tradition, and to the Prophet's sleep. Also, in comparison to the Nocturnal Journey, Dante's voyage starts at dawn: "Temp'era dal principio del mattino,/ E il sol montava 'n su con quelle stelle,/ Ch'eran con lui quando l'Amor divino/ Mosse di prima quelle cose belle" (*Inf.* I, 37), since for the neophyte the dawn announces the change of the cycle and of his state.

Nocturnal Journey,[1] since the spiritual realization, regardless of the traditional form, obeys the same unique Truth.[2] Dante commences his journey at the "North Pole," descends along the "spiritual axis"[3] in Hell, and climbs Mount Purgatory at the opposite end, that is, at the "South Pole." As in the case of the Prophet Muhammad, there is a "double center," or we should say that there is a double "double center," because the starting point is marked by two known symbols of the center: the forest and the city, and also the starting point is, in essence, identical to the arrival point. Dante finds himself wandering in the savage and obscure forest (*in questo centro*), yet at the same time he is placed in the terrestrial Jerusalem (that represents for him, as Guénon said, the "spiritual pole"), which is a prefiguration of the Heavenly Jerusalem; at the opposite pole, there is the Earthly Paradise where, consequent to the spiritual regeneration and the coming of a new "Golden Age," the Heavenly Jerusalem will descend. In fact, there is no "North" Pole and "South" Pole, two different centers, but the one and only Pole, the unique Center, which only apparently the initiate will reach following a journey from center to center, since all these *chakras*, all these various centers are merely projections of the unique Center. Wandering in the middle of the *selva selvaggia*, Dante discovers in front of him a luminous hill, bathed by the rays of

---

[1] For all the minutiae regarding the Nocturnal Journey and a detailed comparison between Dante's voyage and this one see Palacios; see also Ibn 'Arabî, *L'Arbre du Monde*, Les Deux Océans, 1982, pp. 171-176, Ibn 'Arabî, *L'Alchimie*, p. 57. However, the description of the Nocturnal Journey found in the Islamic tradition is, as expected, similar to other spiritual voyages belonging to various traditional forms, including fairy tales. We should mention the well known *Book of Arta Viraf*, where the hero accomplishes a similar journey, ascending by degrees until he reaches Ahura Mazda; Dante's *Divine Comedy* was often compared to this.

[2] In fairy tales, the hero covers the same journey, from center to center.

[3] Guénon, *L'ésotérisme de Dante*, p. 65. Since the Mount Purgatory is situated at the South Pole, Jerusalem should be at the North Pole; concerning the Purgatory's location, Dante said: "Io mi volsi a man destra, e posi mente/ All'altro polo [the other pole], e vidi quattro stele [four stars]" (*Purg.* I, 22-23); the four stars represent, even symbolically, the Southern Cross, and the other pole is the South Pole.

the "intelligible sun," a hill or mountain placed at the "North Pole" and described as "il dilettoso monte,/ Ch'è principio e cagion di tutta gioia"[1]; obviously, Mount Purgatory, situated at the opposite end, at the "South Pole," on an island in the middle of the ocean,[2] is not different from this luminous mountain (*bel monte*[3]), which Dante could not reach directly, but had to follow the "spiritual axis" and descend into Hell.

The absolute Center is presented in both journeys as a super-luminous Light surrounded by cascades of lights. This is a universal symbolism and we should remember what Saint Dionysius the Areopagite said in his Fifth Letter: "The divine darkness is that 'unapproachable light' where God is said to live.[4] And if it is invisible because of a superabundant clarity [the super-luminous Darkness,[5] the supernal Night[6]], if it cannot be approached because of the outpouring of its transcendent gift of light, yet it is here that is found everyone worthy to know God and to look upon Him. And such a one, precisely because he neither sees Him nor knows Him, truly arrives at that which is beyond all seeing and all knowledge." At the same time, it is important to note, Dionysius the Areopagite compared God to an immutable Rock: "Imagine a great shining [super-luminous] chain hanging downward from the heights of heaven to the world below. We grab hold of it with one hand and then another and we seem to be pulling it down toward us. Actually it is already there in the heights and down below and instead of pulling it to us we are being lifted upward to that brilliance above, to the dazzling light of those beams. Or picture ourselves aboard a boat. There are hawsers joining it to some rock. We take hold of them and pull on them, and it is as if we were dragging the rock to us when in fact we are hauling

---

[1] *Inf.* I, 77-8.
[2] Râvana's center, in the *Râmâyana*, is very similarly described, as being situated in the middle of the ocean, on an island, at the top of the mountain.
[3] *Inf.* II, 120.
[4] Saint Paul wrote: "Who alone is immortal and who lives in unapproachable light" (*1 Timothy* 6:16).
[5] *The Mystic Theology*, I, 1.
[6] The "supernal Night" is what Ibn 'Arabî called "nocturnal."

ourselves and our boat toward that rock. And, from another point of view, when someone on the boat pushes away from the rock which is on the shore he will have no effect on the rock, which stands immovable, but will make a space between it and himself, and the more he pushes the greater the space will be."[1]

Prophet Muhammad's *Mi'râj* is marked by the symbolism of light: the super-luminous light was indescribable when, approaching the Throne of God, a green garland brought him to God[2]; the same luminous description is used by Dante[3] for whom the apex of the Center, of the Heavenly Paradise is the Light of lights, *Alto lume, Vivo Lume*: "O Somma Luce che tanti ti levi"[4]; "A quella Luce, cotal si diventa,/ Che, volgersi da Lei per altro aspetto/ È impossibil che mai si consenta"[5]; "O Luce Eterna Eterna che, sola, in Te, sidi,/ Sola T'intendi, e, da Te, intelletta/ E intendente, Te, ami ed arridi!"[6]

René Guénon described the Center as a universal spherical vortex: "If we consider this extension a geometrical symbol, that is, a spatial one, of the total Possibility, we should obtain the representation (insofar this illustration is possible) of the universal spherical vortex, along which flows the realization of

---

[1] *The Divine Names*, III, 1. See Pseudo-Dionysius, *The Complete Works*, Paulist Press, 1987, p. 68.
[2] See Palacios 37, 40, 49. In a *hadîth*, the Prophet affirmed that he saw God as Light. Also, at the end of his own nocturnal journey, Ibn 'Arabî declared: "Then I was enveloped by the divine lights until all of me became Light" (*The Meccan Revelations*, I, pp. 228, 349); Ibn 'Arabî also mentioned, with respect to the Journey, the "Grandiose Light" (*Nûr a'zam*) situated just before the "Throne of Merciful" (*L'Alchimie*, p. 125).
[3] Palacios 46, 50, 51. The architects and Masons of the Gothic cathedrals perfectly understood the symbolism of light. From the beginning of the *Paradiso*, Dante indicated: "E, di subito, parve giorno, a giorno,/ Esser aggiunto, come Quei, che puote,/ Avesse il ciel, d'un altro sole, adorno./ Beatrice, tutta, nell'eterne rote,/ Fissa con gli occhi, stava; ed io, in lei,/ Le luci fissi, di lassù, remote" (*Parad.* I, 61); the Principle is called *la Verace Luce* (*Parad.* III, 32), the eternal Light ("la Luce, che sola sè medesma vede compiutamente" *Conviv.*), residing in itself.
[4] *Parad.* XXXIII, 67.
[5] *Parad.* XXXIII, 100.
[6] *Parad.* XXXIII, 124.

all things, and which the metaphysical tradition of the Far-East calls Dao, that is, the 'Way.'"¹ Ezekiel, we noticed already, had also a vision of God as universal vortex, a "world machine" of fire and lightning. The same vision is present in the *Mi'râj* and Dante's *Paradiso*: there is a luminous and holy vortex, composed of numerous hierarchical circles,² rotating around God.³

If in the *Mi'râj* the Throne of God is the Great Extreme, there are other symbols of the Center, like *Bayt-al-Ma'mûr* ("the Much-Frequented House"), which, it is said, Allâh has built it from a red ruby stone, while its two opposite doors were made of green emerald⁴; during his Ascension, the Prophet Muhammad visited the House (called also the "Inhabited House"), which is illuminated by ten thousand lamps of pure gold and seventy thousand angels come to visit it, entering through one door and exiting through the other.⁵ Ibn 'Arabî

---

¹ Guénon, *Le symbolisme de la croix*, p. 112.
² About the paradisiacal hierarchy, see the sayings of the greatest spiritual master Ibn 'Arabî, who, similar to the Hindu tradition, explained that each individual resides the degree corresponding to his stage of knowledge (and being unaware of this hierarchy, since each one sees God in a "form" that is in accord with his knowledge), all obeying the law of harmony and perfect order. See Palacios 260-1, 267, 269, 273, 274; *The Meccan Revelations*, I, p. 117. "The vision of God on the Day of the Visit is according to men's beliefs in this world. Thus the person ... sees his Lord in the form of the aspect of each belief he held concerning Him" (*ibid.* p. 118). See also Ibn 'Arabî, *L'Alchimie*, p. 59.
³ See Palacios 38, 61, 275-278. Ibn 'Arabî used the geometrical symbolism (circles) to express the same reality.
⁴ Or it is described as "a building on four pillars of emerald, which he crowned with a hyacinth, and this building was called al-Durâh" (Wensinck, *The Navel*, p. 50).
⁵ "Abraham commands the adept to enter in the Inhabited House [called also the 'blissful abode' or the 'Temple inhabited with believers'] ... Then, he walks out through the same door he entered, instead of leaving through the "angels' gate" (*bâb al-malâ'ika*) which is the second door of the sanctuary. This one has a special characteristic: the one who goes out through this gate cannot come back" (Ibn 'Arabî, *L'Alchimie*, pp. 105-6). The first door is, consequently, the "men's gate," the gate that allows a return, and we could compare these two gates with the Hindu *dêva-yâna* and *pitri-yâna*, but here the "return" refers to the complete spiritual realization. See also Gilis, *Pèlerinage*, p. 46.

considered the House to be situated in the seventh heaven,[1] but for him, since the Nocturnal Journey symbolized a complete spiritual realization, the Inhabited House has the heart as an equivalent in man.[2] Similar to Nicholas of Cusa, Ibn 'Arabî affirmed: "God placed the level of the Universal Man midway between God's being seated upon His Throne and His being located within his heart, which encompasses Him. So he looks upon Him in his heart and sees that He is the Central Point of the Circle; and he looks upon Him seated upon His Throne and sees that He is the Circumference of the Circle."[3] Ultimately, the Heart is the absolute Center containing all the places and events of the Journey, either heavens, prophets and angels, or the Inhabited House and the Throne, since the stages of the spiritual realization are projections and images of the unique Center, which is God.[4] And Ibn 'Arabî insists in defining the Nocturnal Journey as an inner initiatory voyage within the heart, since, according to a *hadîth qudsî*, "My earth does not encompass Me, nor does My heaven, but the heart of My servant, the man of true faith, does encompass Me." Moreover, Ibn 'Arabî adamantly stressed that the Nocturnal Journey was not only a *mi'râj*, an ascension, but also a "return" (*rujû'*), since only so is the spiritual realization completed, a "return" characterizing the highest rank of the saints and prophets.[5]

---

[1] Palacios suggested that the House is in the eighth circle or just under the Throne of God (p. 24).
[2] *The Meccan Revelations*, I, p. 38. Ibn 'Arabî described his own nocturnal journey: "Then I saw the Inhabited House, and suddenly there was my Heart" (p. 228).
[3] *The Meccan Revelations*, I, p. 188. And Ibn 'Arabî added: "His Circumference is His Names, while His Center is His Essence.
[4] "This Kaaba of Mine is the Heart of being, and My Throne (the whole universe) is a limited body for this Heart," Ibn 'Arabî said (Morris, *The Reflective Heart*, p. 56).
[5] *The Meccan Revelations*, I, pp. 314, 323. However, we can speak about a "return" only with regard to the beings residing in manifestation (like the human beings for example), because for them nothing is changed and the hierarchy of the states still exists, while for the Universal Man the *Axis Mundi* was absorbed into the unique point, which is the supreme Center (René Guénon, *La Grande Triade*, pp. 209-210). Also, the "one who returns" has to

René Guénon, in a fundamental article, *Réalisation ascendante et réalisation descendante*,[1] unveiled an essential characteristic of the initiatory process: the total spiritual realization implies, beside the ascendant journey, a descendant phase,[2] since only when both the non-manifestation and manifestation are integrated (even though the second domain is negligible with respect to the first one) is the Universal Man truly and fully realized.[3] In the Islamic tradition, Guénon said, the difference between the two phases is expressed, with a good approximation, by the two functions: *walî* (the seer for himself, who accomplished the ascendant journey) and *nabî* (the spiritual master for others).[4]

---

simulate a common appearance to accomplish the "return"; otherwise, as in the case of Moses, the inferior levels are not integrated, and the regular people cannot participate (to the best of their abilities) to the mystery of the transcendent light; this common appearance, which could be defined as "obscure," is, for the "one who returns," the image of the supernal darkness (the "super-luminous tenebrae") (Guénon, *Initiation et réalisation spirituelle*, pp. 219, 227; see below Ibn 'Arabî about Moses).

[1] Guénon, *Initiation et réalisation spirituelle*, pp. 251 ff.
[2] This second phase could be found in many initiatory fairy tales.
[3] Guénon quoted the *Tabula Smaragdina*: "It ascends from Earth to Heaven, and descends again to the Earth, taking unto itself thereby the power of the Above and the Below"; from an initiatory viewpoint, this quotation illustrates the double realization, "ascendant" and "descendant" (*La Grande Triade*, p. 120). See also about this Michel Vâlsan, *Les derniers hauts grades de l'Écossisme et la réalisation descendante*, Études Traditionnelles, no. 308, 309, 310, 1953. In fact, Michel Vâlsan already published in a previous issue (no. 307, 1953), an article called *Un texte du Cheikh el-akbar sur la «réalisation descendante»*.
[4] Following St. Thomas Aquinas, we could say that Love for God (*bhakti-mârga*) represents the ascendant phase, and the Mercy or Compassion is the descendant one. In Islam, the function of *rasûl* has, in comparison with that of *nabî*, a universal character and manifests in all the worlds the divine attribute of *Ar-Rahmân* (The Most Merciful); see also Ibn 'Arabî in *The Meccan Revelations*, I, pp. 218-219. As Michel Vâlsan said, the descendant realization implies the intervention of an angel, like in the case of a *rasûl* or a *nabî*, since God chooses one on whom to confer a "mission," which represents precisely the descendant initiation. The divine "missionary" becomes a "victim," since to leave the supreme center is a "sacrifice" ("the night is better than the day," Guénon, *Aperçus sur l'initiation*, p. 156, *Symboles fondamentaux*, p. 243); therefore, Vishnu had to be convinced to come down as an *avatâra* (see Guénon, *Initiation et réalisation spirituelle*, pp. 264-5, where the descendant realization is

Guénon and Vâlsan are in concert with Ibn 'Arabî, for whom the "return" is even more important than the ascension, albeit almost nobody insists on it; for the greatest spiritual master, the "true certainty" (*yaqîn haqq*) (which is the highest stage, above the "certainty of knowledge" and the "certainty of seeing") is reached only when the "return" is accomplished as well, since only then "every property [of reality] becomes firmly established in its proper rank, and things are not confused for him." This "return" is also about accepting the whole circle – center, circumference, and radii – and recognizing that the human factor has its role, which cannot be wiped out, in the same way as in the Hindu tradition, where the realization of the Self (*Âtmâ*) does not actually mean the extinction of the *ego*, but its transformation.[1]

---

related to the idea of "mission" and *avatâra*, since all the divine "missionaries" are achieving a descent).

[1] *The Meccan Revelations*, I, pp. 110-111. As James W. Morris noted, Ibn 'Arabî stressed that *fanâ* and the "divine vision" should not be considered the ultimate stage of spiritual realization, since the final stage of perfection requires a process of integrative "return" to the world that finally allows the Knower to perceive all things "as they really are." It is a lack of "maturity" to deny the reality of this world, as some low rank "knowers" do, in contrast to the true "heirs" of the prophets who are always aware of God's Presence throughout this world (p. 340). Ibn 'Arabî described this complete realization (ascendant and descendant) insisting upon the reintegration of the being, including the *ego* (this symbolism can be decrypted in the initiatory fairy tales): during the ascension, the initiate leaves in each world that part of himself which corresponds to it; during the descent, he takes from each world that aspect of himself which he has left there and reintegrates it in his self, until he arrives back on earth (pp. 212-214). For Ibn 'Arabî, this is the difference between the Prophet Muhammad and Moses, since the first, after his return, did not show any signs indicating his Ascension (as a result of the complete realization), while the second, after he came back from the Mount Sinai, was shining with the divine light (*Exodus* 34:29) (p. 329). See also *Journey to the Lord of Power*, pp. 51-52, 59. On the other hand, as Guénon explained, in the *Kundalinî-Yoga*, the ascension of the *Kundalinî*, through the *sushumnâ*, means a journey from center to center, and at the each reached center *Kundalinî* reabsorbs the various principles of the individual manifestation, bringing them to a potential state, and in this state they will be carried to the next superior center (*Études sur l'hindouisme*, p. 37). It seems that *Kundalinî-Yoga* would contradict Ibn 'Arabî, but that is not the case, since what the initiate leaves in

With respect to the "human factor" it is important to underline that Ibn 'Arabî sees its validity only as God's manifestation and not as an independent factor, and the realization of the "return" is genuine only if we "read" into humankind God's Signs.[1] As Ibn 'Arabî stressed numerous times, the spiritual realization aims at comprehending God's Signs, and that includes those of our world considered as God's trials.[2] We ought to add that Ibn 'Arabî, like Guénon later, explained how the devil is, without knowing, following the divine commandment and "ends up accomplishing the exact opposite of what he intended" as his deception is a "trial"

---

his ascension are the clothes (the "explication"). From the Hindu tradition's point of view, we could say that during the ascendant realization the various envelopes of the *Âtmâ* are peeled off and this is the quest for the center; when the center is achieved, it becomes evident that, in fact, the *Âtmâ* contains all the envelopes, as the *Spiritus* contains the *Corpus* (see Guénon, *Initiation et réalisation spirituelle*, p. 233), and during the descendant realization the envelopes, which are gradually reintegrated, appear to the liberated being as they really are, that is, comprised and enveloped by the *Âtmâ* (in fact, now is fulfilled the saying of the *awliyâ*: "our bodies are our spirits [*Spiritus*], and our spirits are our bodies"; *Initiation et réalisation spirituelle*, p. 238).

[1] The *Qur'an* (and Ibn 'Arabî) stresses the crucial meaning of "God's Signs"; in the Christian tradition, we learn: "And as Jesus passed by, he saw a man which was blind from his birth. And his disciples asked him, saying, Master, who did sin, this man, or his parents, that he was born blind? Jesus answered, Neither hath this man sinned, nor his parents: but that the works of God should be made manifest in him. I must work the works of him that sent me, while it is day: the night cometh, when no man can work. As long as I am in the world, I am the light of the world" (*John* 9:1-5).

[2] See *The Reflective Heart*, pp. 249, 300. "We will try you with something of fear, and hunger and loss of wealth, and souls and fruit; but give good tidings to the patient, who when there falls on them a calamity say, Verily, we are God's and, verily, to Him do we return" (*Qur'an* 2:155-156). Yet we must not confuse the exoteric with the esoteric; as René Guénon pointed out, the "life's trials" are not initiatory trials, and the latter are in fact rites of purification (*Aperçus sur l'initiation*, p. 172 ff.); Ibn 'Arabî said: "Now spiritual purification – which is (purification of) the heart – is through liberating ourselves (from all attachments other than God), in order to seek (His) friendship. And there is no (true) friendship and closeness with God except through freeing yourself from the creatures, insofar as you used to consider them (only) in light of their relation to yourself (to your ego or *nafs*) and not through God (and the realization of His aims in their regard)" (*The Reflective Heart*, p. 88).

pushing the man to regret and then to true repentance and a return to God; Ibn 'Arabî says that "this is the divine Cunning (*makar Allâh*) through which He fools Iblîs."[1]

In his description of the Ascension, Ibn 'Arabî parted it into three stages: the first phase with the seven heavens and the Inhabited House in the seventh; the second one with seven steps, from the Lotus-Tree of the Limit to the Throne of the Merciful[2]; and the third phase leading to the Divine Presence of the One.[3] Occasionally, Ibn 'Arabî called the Inhabited House

---

[1] *The Reflective Heart*, pp. 92-93. There is another divine Cunningness which we should mention for the record: "And they (the Jews) devised a stratagem against him [Jesus], and Allâh schemed (against them): and Allâh is the best of schemers" (*Qur'an* 3:54); "And because of their saying: We slew the Messiah, Jesus son of Mary, Allah's messenger – they slew him not nor crucified him, but he was represented by one in his likeness; and lo! those who disagree concerning it are in doubt thereof; they have no knowledge thereof save pursuit of a conjecture; they slew him not for certain" (4:155-157). The similarity was noticed between the above text, the Docetism and some Gnostic Gospels (where Judas is considered the twin brother of Jesus), yet we should remember Michel Vâlsan's opinion that, generally, those who talked about Docetism did not really understand it, and from an initiatory point of view there were very interesting elements in Docetism (Michel Vâlsan, *Références islamiques du "Symbolisme de la Croix"*, Études Traditionnelles, no. 428, 1971, p. 278). See also Robert de Boron, *Merlin and the Grail*, D. S. Brewer, 2007, pp. 16-17, where the Judas' kiss is explained as a way to differentiate Jesus from his cousin James the Lesser, who "looked very much like Jesus"; we may note that the Grail stories transmitted an esoteric lore. In the end, we should mention Nicholas of Cusa, who considered that the Muslims negated Jesus' crucifixion because they respected Christ and for them the Jews had no power upon Him (*La paix de la foi*, p. 74).

[2] Each represents a center.

[3] Ibn 'Arabî, *L'Alchimie*, pp. 109, 131. Ibn 'Arabî separated as well the Nocturnal Journey with respect to the celestial mounts participating in it: Burâq, the "lightning," from Mecca to Jerusalem; the second mount, from Jerusalem to the first heaven; the third, represented by the angels' wings, from the first to the seventh heaven; the fourth, represented by Gabriel's wings, from the seventh heaven to the Lotus-Tree of the Limit; the fifth, "the carpet deployed like a flying bird," Rafraf, from the Lotus-Tree to the Throne; the sixth mount, al-Ta'yîd, from the Throne to the Divine Proximity (see Ibn 'Arabî, *L'Arbre du Monde*, pp. 99-102; about the center called the Divine Proximity, see also *The Meccan Revelations*, II, pp. 232-3); the celestial mounts,

the "Marvellous City" (*al-Madînat al-fâdila*), which could be an equivalent to the Heavenly Jerusalem,[1] since, sometimes, the Inhabited House of the seventh heaven was considered to be the Temple of the Heavenly Jerusalem and was often identified to the *Al-Masjid El-Aqsa* of the *Mi'râj*.[2] The Much-Frequented House stands, as a *hadîth* says, right above the terrestrial Kaaba and if it had fallen it would fall right on top of the Kaaba.[3]

We should remark first that, essentially, *Bayt-al-Ma'mûr* is a stone (a "red ruby") and so is the Kaaba, the cubic form alluding to a perfect stone, what the Masons call *ashlar*, which is directly related to the symbolism of the rock as center. The Heavenly Jerusalem is also a stone : "that great city, the holy Jerusalem, descending out of heaven from God, having the glory of God: and her light was like unto a stone most precious, even like a jasper stone, clear as crystal"[4]; and a cube: "The length and the breadth and the height of it are equal."[5]

Second, the "spiritual axis" or the "vertical" one[6] connects the earthly center to the heavenly center – a truth found in all the traditional forms. Considering that the Ascension started in Jerusalem (the terrestrial secondary center), the spiritual traveler who follows this "vertical" direction will reach the Heavenly Jerusalem, which is situated right above.[7] What we have here is

---

which contain a rich initiatory symbolism, are in direct relation with these various centers.

[1] Ibn 'Arabî, *L'Alchimie*, p. 36.
[2] Ibn 'Arabî, *L'Alchimie*, pp. 23, 98. See also Palacios 24, where it is said, following a famous *hadîth*, that Abraham was leaning his back to the temple's wall of the Heavenly Jerusalem, identical to the Inhabited House.
[3] See also Wensinck, *The Navel*, p. 48.
[4] *Revelation* 21:10-11.
[5] *Revelation* 21:16. The Holy of Holies in the Temple of Jerusalem is also a cube (Burckhardt, *Art of Islam*, p. 4).
[6] See Guénon, *Le symbolisme de la croix*, p. 123.
[7] Palacios quoted Ka'ab al-Ahbâr (the Jewish Rabbi from Yemen, who accompanied Caliph Omar on his voyage to Jerusalem, *Al-Quds*, and, it is said, helped locate the foundations of the ancient Temple and the place of the Rock): "The Paradise is in the seventh heaven, in front of Jerusalem and of the Rock; if a stone had fallen it would fall right on the Rock." Palacios quoted as well a medieval Islamic text (the 11th Century): "Jerusalem is the

obviously the symbolism of the "double center" and, since the Heavenly Jerusalem is the Center of our world, as Guénon said, *Bayt-al-Ma'mûr* should be no different from it.

The symbolism of the pair Heavenly Center – Earthly Center was well known in all the traditional forms, as we can observe in the case of Mesopotamia, where the traditional civilizations that flourished there had many affinities to the other Semitic societies. Gudea, who ruled in the city of Lagash two thousand years before Christ, saw in a dream a god with a tablet containing the plan of the temple he wanted to build; Sennacherib, when he founded Nineveh, as a center of his kingdom, said that the city "was planned from far-away time in the writing of heaven"; not to say that each city or temple had the corresponding constellation in heaven.[1] The Judaic tradition underlines: "The Lord is in his holy temple, the Lord's throne is in heaven"[2]; and the Holy of Holies below is situated parallel to the Holy of Holies above. As we know, the Tabernacle was built "according unto the pattern which the Lord had shewed Moses"[3]; with regard to the Temple, "all this, said David, the Lord made me understand in writing by his hand upon me, even all the works of this pattern."[4] Similarly, as we already saw, the earthly Kaaba has its archetype in the heavens; "at the beginning," Michel Vâlsan wrote, "before Adam's descent,[5] the Kaaba sanctuary was, in a primary form, a center for the angels' pilgrimage," which means that Kaaba belonged to the informal manifestation, being identical to the celestial pole.

---

navel of the Earth. The gate of heaven is open above its Temple ... God said to the rock of the Temple: You are my inferior throne. From you, the heaven ascends to Me" (p. 236). Ka'ab al-Ahbâr also said: "Allâh says to Jerusalem: Thou art my throne, from which I ascended to heaven" (Wensinck, *The Navel*, p. 24).

[1] See *The Labyrinth*, pp. 60-1, 64-5.
[2] *Psalms* 11:4.
[3] *Numbers* 8:4.
[4] *1 Chronicles* 28:19.
[5] Ibn 'Arabî stated that, in fact, the fall of Adam "was a descent manifesting his sovereignty and his viceroyal function, and not a banishing."

As the first projection of the Heavenly Temple, the Kaaba designates the Earthly Paradise, where the visiting angels symbolize the spiritual influences, and Adam's descent imposed the projection of the Temple into the world as its center.[1]

Adam's descent is accompanied by paradisiacal (divine) things, symbolizing the revealed Tradition: the *Tâbût*,[2] the *Rukn*, and the stick (*al-Asâ*)[3]; the *Rukn* is the "angle-stone" that became the black stone of Kaaba.[4] Ibn 'Arabî said: "The House of Allâh differs from the Throne (*ad-Durâh*) and from the other fourteen 'houses' by a thing that no tradition attributes it to those, which is the black stone, the Right Hand of Allâh on earth."[5] The black stone is an equivalent of the Rock as center, even though it is placed in a "corner" (*rukn*) of the Kaaba, since the center (Burckhardt considers) should remain somehow invisible, non-located, in accord with the spatial expansion of the nomads.[6] Ibn Abbâs affirmed: "The black stone that descended from Paradise [like the Grail stone] was whiter than milk; then, the sins of Adam's sons blackened it (*sawwadat-hu*)."

---

[1] A *hadîth* says: "Before Adam's descent from Paradise, the House (the Temple, *al-Bayt*, the Kaaba), was a hyacinth among the hyacinths of Paradise."

[2] We already mentioned the *Tâbût* as a chest or ark – in fact a primordial and *principially* everlasting repository, "Adam's treasure," the Holy Grail, the Tradition. The Islamic tradition mentions the "treasure" in the Kaaba, which is in reality Allâh (Gilis, *Pèlerinage*, pp. 47-49); similarly, the Templars' "treasure" is Christ, a truth impossible to be understood by the modern mentality.

[3] Gilis, *Pèlerinage*, pp. 66, 68. These three heavenly objects correspond in a way to the three universal functions. The stick belonged to Adam and was inherited by Moses. There is an interesting parallel between this Islamic tradition and the initiatory fairy tales, where such divine objects are always present.

[4] About the *rukn* see also Guénon, *Symboles fondamentaux*, pp. 282-3. "The black stone is the right hand of Allâh on earth" (Gaudefroy-Demombynes 46). The sacred rock of *Qubbat As-Sakhrah*, Corbin specified, has a function homologous to that of the black stone (Henry Corbin, *Temple et Contemplation*, Flammarion, 1980, p. 391).

[5] We may note Ibn 'Arabî's indication about the existence of many centers, one for each "level." See also Burckhardt, *Art of Islam*, p. 4, Gilis, *Pèlerinage*, p. 46.

[6] Burckhardt, *Principes*, p. 147.

In fact, this explanation is mostly exoteric, since from a metaphysical point of view, the Center is both white and black, and so should the stone of Kaaba be[1]; the black-white stone is, essentially, the manifestation of the Center as temple, the Islamic tradition reporting how it descended from Heaven.[2]

We should note that Adam descended in India (which symbolizes the land of the Primordial Tradition), and from there he commenced his initiatory "quest of the House" with the intention of founding a secondary center and a new cycle.[3] Adam established the House of the new tradition in the shape of a Tabernacle or Tent (*khayma*),[4] having as its archetype the heavenly Kaaba – a sacred deed Adam was compelled to accomplish because of the cycle's decadence, when the people could not have direct access anymore to the heavenly mysteries and it was necessary to have these mysteries descended among the people and to found a new traditional form capable of teaching them again the way towards the center.[5] The Tent was a circular house supported, like the divine Throne, by four

---

[1] About the supernal sense of the colour black see, for example, Guénon, *Symboles fondamentaux*, pp. 306-308. Guénon mentioned the Masonic *tessellated pavement* and the Templars' *Beaucéant*, which are too black and white.

[2] The stone that descended from Heaven is the *lapsit exillis*, about which see Guénon, *Symboles fondamentaux*, p. 292. There is a direct relation between the "black stone" and the "white pearl" (about this white pearl, in the Judaic Kabbalah and at Gnostics, see our *The Wrath of Gods*, pp. 145 ff.). In Shi'ism, it is said that the stone was, at the beginning, an angel among angels, whom God planted, beside Adam, in Paradise; when Adam descended to earth, God transformed the angel into a white pearl that He projected from Paradise to Adam, in India, but he did not recognize it (there is here a similarity with the Gnostic myth), until the pearl reminded him about his covenant with God. After that, God transformed the pearl into the black stone, which Adam carried with him from India to Mecca (Corbin 256-257).

[3] Mecca appears as a secondary center also in comparison to Mount Arafa (the Mount of Mercy, *Jabal ar-Rahmah*) (*Qur'an* 2:198), where Adam met and knew his wife after a long separation (Gilis, *Pèlerinage*, p. 70).

[4] This tent was a red hyacinth (or was made of a red ruby, Gaudefroy-Demombynes 30). See also Gilis, *Pèlerinage*, p. 71.

[5] Even in this case we encounter the symbolism of the double center, since a *hadîth* says that Adam, forty years after he built the Kaaba, built the *Al-Aqsa* Mosque.

columns made of precious stones or by four white *arkân*; it is also said that the Tent was self-illuminated, with a dome (*qubba*) of red hyacinth and four doors (oriented towards the cardinal points) corresponding to Adam, Abraham, Ishmael, and Muhammad; another tradition described the primordial Kaaba built by Seth, Adam's son, in a shape of a pyramid, which was destroyed by the flood[1]; Abraham rebuilt it in the shape of a cube (*ka'bah*).[2]

The change from pyramid to cube meant an important modification of the tradition and, considering that this change occurred after the flood, we could assume that this readaptation took place after the disappearance of Atlantis, the cubic form suggesting the last phase of the cycle. Abraham, as Eber's successor (Eber being Sem's grand-grandchild), was the father of the Semitic "eberits," Hebrews and Arabs, yet, as Tha'labî said, Abraham was considered *abrani* (Hebrew) and Ishmael *arabi* (Arab).[3] Abraham's migration towards the West represents a Chaldean current[4] – the junction point of two major

---

[1] Burckhardt, *Principes*, p. 29.

[2] In conformity to another tradition, the celestial archetype of the Temple appeared in two forms: as Tent descended from Heaven and as a white Cloud (Corbin 232). In the first case, because Adam complained about the loneliness and wilderness of the place, God lowered a Tent from the paradisiacal tents; the surface of the tent covered precisely that of the future sanctuary, and the central pillar was made of red hyacinth. Four stones, from Mount Safâ, Mount Sinai, Mount Salem and Mount Abû Qubays, were used as foundation-stones. In the second case, the archangel Gabriel guided Adam to the sacred place of the Temple (since not any place could be a secondary center), and a white Cloud (*ghamâma*) covered it with its shadow (Corbin 240).

[3] Gilis, *Pèlerinage*, p. 81.

[4] As Michel Vâlsan said, Adam's "stations" ("houses"), the change of his stature (he was a giant who reached the heaven with his head and, following the angels' complaints, he was reduced to a common man's height, which implies two phases, a divine and a human one), and his attitude, all envisage in fact a "humanity" or a cycle, and not an individual; Adam's journey, as pilgrim from India to Mecca, could signify the displacement of an intellectual aggregate, of a spiritual current and of a human group. In a similar mode we could fathom some elements of Abraham's history corresponding to a current associated with Hinduism, while Ishmael relates to Islam (Gilis, *Pèlerinage*, pp. 69, 76). Adam's colossal stature (illustrating the Universal Man) is similar to

traditions, the Hyperborean and the Atlantean one, and the primary source of the Judaic tradition. During this turmoil, there will emerge as well an "Ishmaelian" tradition, resulting from the fusion of the Abrahamic and Egyptian currents,[1] from where the Islamic tradition will develop.[2] Returning from Egypt, Abraham received Melchizedec's blessing, which marks, as Guénon said, the junction point between the Hebraic tradition and the great Primordial Tradition;[3] then, Hagar gave birth to Ishmael[4] and, after that, Abram "changed essentially" into Abraham.[5]

Since, in the building of the Kaaba, Abraham has the active role and Ishmael the passive one (he is the "listener"), Michel

---

that of the Divine Being from the *Râmâyana*, whose height was like a mountain peak and who brought down the celestial *Pâyasa*.

[1] Ishmael's mother and wife were Egyptians.

[2] Abraham facilitated the encounter between the Sumerian and Egyptian civilizations. We should also remember, as Guénon said, that the Judaic tradition was modified and finalized in Egypt, where it received the Atlantean influence.

[3] *Le Roi*, p. 50.

[4] Ishmael's name is related to the spiritual act of "listening." It is not only about the primordial sound (*Parashabda* in the Hindu tradition) and the fact that the Tradition was "heard" (*shruti*), but it is also about *islam*, the total servitude with respect to God; moreover, God listens to Ishmael's cry. Hagar, on the other hand, is connected to the idea of seeing (*Genesis* 16:11, 16:13-14, 21:15-18). As Tha'labi narrated, Hagar is also directly linked to the "sound," since she "heard" a voice that made her run between Safâ and Marwa (this race, which is part of the pilgrimage to Mecca, is present also in some initiatory fairy tales, since running has an important symbolic meaning). Ishmael, who is a "chosen" one, a divine "orphan," marks the beginning of a new cycle, related to the sound and the center (as Tha'labi said, Hagar and Ishmael were left at *hijr* in Mecca); in various traditions the "orphan" is linked to the future temple. About listening and seeing see the sayings of Ibn 'Arabî in Morris, *The Reflective Heart*.

[5] *Le Roi*, p. 48. All these illustrate various adaptations preparing new traditional forms. Ishmael was considered a "wild man" (*Genesis* 16:12) (in fact, *pere' âdâm*, where *pere'* = "wild ass"), a hunter like Nimrod, which suggests a *Kshatriya* nature (about the relation between the "Nimrodian spirit," the Egyptian Set, and the "wild ass" see Guénon, *Symboles fondamentaux*, pp. 47-8). The Islamic tradition insists on describing Abraham's function, which was to affirm the pure spirituality and combat Nimrod's illegitimate ambitions.

Vâlsan considered that Abraham corresponds to the inner, silent Word (The Thought, the superior semicircle), and Ishmael to the exterior, pronounced, manifested Word (the inferior semicircle); therefore, Abraham and Ishmael, together, illustrate the center as the World's Egg.[1]

The tradition reports that the divine Presence (*sakînah*), manifested as serpent, led Abraham to the sacred place where the Kaaba had to be built, which illustrates the idea of the center manifested as temple, followed by that of the temple manifested as center.[2] It is also said that Allâh, since Abraham did not know how to build the House, sent down *Sakînah*, as a whirlwind with two heads [the two serpents of the Caduceus, or the pair Seth-Set] to point out the location of the sanctuary, or as a cloud, which guided Abraham (who is a true Masonic architect[3]) in building the temple, giving him the right measurements (while Ishmael is not mentioned). It is said that Abraham built the temple as center using the stones brought by the angels from seven mountains, as a symbol of his function – the heir of the wisdom of the previous cycles and the adaptor of the Tradition to the new cyclic conditions. At the end of the Work, there was left an empty place for the last stone, the black stone that, in a way, is equivalent to the "keystone," which must be found through a spiritual quest; even though Ishmael tried to find it, Tirmidhi stated that Mount Abu Qubays (situated initially in Khorassan[4]) came to Mecca and, by Allâh's permission, offered to Abraham the black stone that had been

---

[1] See Gilis, *Pèlerinage*, pp. 75-76, 79. We could say that Abraham corresponds to the spiritual authority (*Brâhmana*) and the masculine principle (and the Primordial Tradition, expressed as the Hindu tradition) and Hagar-Ishmael to the temporal power (*Kshatriya*) and the feminine principle (and Islam). Due to his passive role, Ishmael is not the one who finds the black stone. See Gilis, *Pèlerinage*, p. 80.
[2] See Gaudefroy-Demombynes 31, Wensinck, *The Navel*, pp. 60, 63, 65, Gilis, *Pèlerinage*, p. 83.
[3] See Gilis, *Pèlerinage*, pp. 85-86.
[4] Khorassan was the region where the Nestorians lived, and their mysterious role in the beginning of Islam is well known (see JeanTourniac, *Lumière d'Orient*, Dervy, 1979).

entrusted by Noah to the mountain to be its repository. And when Ishmael wanted to know how the black stone came into Abraham's hands, this one specified: "The one who gave it to me did not delegate me to tell you," which means that it is Abraham, not Ishmael,[1] who received the Tradition, relating him to Noah and Adam and confirming the Kaaba as a secondary center; Ishmael remains the guardian of the tradition that will become manifest with Muhammad. The birth of the Islamic tradition coincides, as expected, with a rebuilding of the Kaaba, without the modification of the cubic form.[2] Muhammad, in comparison to Ishmael, plays a main and active role; when he was 35 years old and "the divine inspiration did not yet descend upon him," he carried, dressed with a spotted mantle, the stones, beside the Qurayshits,[3] and it was he who has put the last stone, that is, the black stone in its place, accomplishing the construction of the Temple, since Muhammad was the one who received the revelation of the tradition, the black stone being the "trustful repository" (*amâna*), an equivalent of the Tâbût.[4]

---

[1] The stone brought by Ishmael constituted the "station of Abraham" (*maqâm Ibrâhim*).

[2] There are three episodes depicting the same moment of the cycle – the establishment of the Islamic tradition: the fight between the eagle and the snake; the "Peace of Hudaybiyya"; the annihilation of the idols encircling the Kaaba and the inauguration of the Muhammadian Kaaba. All three express the investiture of Muhammad as the regent of the center.

[3] This spotted mantle (*namira*, word related to Nimrod and "tiger") was fabricated of white and black patches (similar to Joseph's mantle). The central symbolism of white and black is stressed too by the fact that the old sanctuary was guarded by *Sakînah*, in the form of a serpent with black back and white front, that was seized by a black and white bird (an eagle like a stork; the symbolism of the battle between the snake and the eagle is well know, as well the symbolism of the stork). As Guénon explained, in the Islamic esotericism, the Great Peace is called *Es-Sakînah*, corresponding to the *Shekinah*, the Divine Presence of the Judaic tradition.

[4] In a way, Muhammad's act could be considered a sacrificial act, typical for such a construction, where Muhammad is identical to the black stone, the same way Zeus is identical to the stone swallowed by his father and which became the Omphalos. Also, as Michel Vâlsan specified, Muhammad could be considered the Architect, since his name al-Amîn seems similar to Amon,

The Rock

There are some traditional data that suggest an esoteric intervention of Christian elements, since it is said that the Qurayshits used the wood transported by a Byzantine ship at the sanctuary's reconstruction[1]; it is also said that the Qurayshits hired foreign craftsmen (like Solomon, because of the nomadic life of the Hebrews and Arabs), called *rûmi* (Byzantines) or *qupti* (Copts), among them Pacôm being especially described as *bannâ* (mason, architect) who contributed to the building of the roof, and as the one who enveloped and veiled the Temple.[2] Let us note that the Prophet Muhammad is the Center, since he is not only an equivalent of the black stone but also of the Kaaba itself, illustrated by the fact that he also veiled and then re-veiled himself again.[3] As René Guénon explained, "to re-veil" means not only to cover again but also to reveal, and the "re-veiling" of Muhammad and of the sanctuary signified, on the one hand, an occultation due to the decadence of the world, and, on the other hand, a revelation.

There is a "violent" aspect of the instatement of this secondary center,[4] related to the fact that the Qurayshits opposed vehemently the new tradition, even though they participated beside Muhammad in the construction of the Kaaba.[5] What we witness here is the uninterrupted transmission

---

another name for Hiram, the architect of Solomon's Temple (*amon* means "architect" or craftsman) (Gilis, *Pèlerinage*, pp. 92-93).

[1] The influence of the Byzantine Empire upon Islam is well known (see our *The Wrath of Gods*); yet what we should understand under the name of "Byzantium" is another question, much more enigmatic. Gilis considered this Christian influence as an esoteric one (*Pèlerinage*, pp. 95-96).

[2] Gaudefroy-Demombynes 33-34.

[3] Gilis, *Pèlerinage*, p. 106.

[4] We know that Ishmael was considered a "wild man."

[5] "Then, when the sacred months have passed, slay the idolaters wherever ye find them, and take them (captive), and besiege them, and prepare for them each ambush" (*Qur'an* 9:5); "Will ye not fight a folk who broke their solemn pledges, and purposed to drive out the messenger and did attack you first?" (9:13). This "violent" aspect is also illustrated by the fact that the first three Caliphs were assassinated and so were Ali's children; however, we should not forget the immense difference between the "lesser war" and the "greater war." Christianity, on the other hand, was under the sign of sacrifice, since not only

of the tradition, guarded by the "Abrahamic" serpent and by the "heirs" of Abraham (Ibrâhim), the Qurayshits, a tradition which was absorbed by the new one, represented by the eagle (or the bird)[1]. With Islam, the "angle" of the Kaaba was not the *hijr* anymore, but the angle of the black stone, which was oriented towards Arafa – a spiritual orientation towards the Heart.[2] However, as we said, the new Kaaba was not only "veiled" but also "revealed," since, even though the sanctuary was "hidden" and "covered," it was at the same time "discovered," exteriorizing the invisible temple, that is, the *hijr*, which became accessible to the people at large.[3]

In a way, the *hijr* has its analogous projection in the *mihrâb*, the semicircular niche of each mosque, which indicates the *qiblah*. As René Guénon explained, the temple as Heaven and

---

Christ, but the majority of the Apostles and many saints died as martyrs (yet not "fighting"); nonetheless, there is too a direct allusion to the "greater war" when Christ affirmed: "think not that I am come to send peace on earth: I came not to send peace, but a sword" (*Matthew* 10:34); and this "violent" aspect was present in the times of the Crusaders.

[1] Despite the "Peace of Hudaybiyya," Muhammad was forced, in 630, to conquer Mecca. The Qurayshits represent the "old" cycle, as the revolt of the Qurayshit Ibn Zubayr against the Omayyads proves it; we may note that Ibn Zubayr tried to rebuild the Kaaba (destroyed by fire) in its ante-Islamic form, including the *hijr* or *hatîm*. Ibn Zubayr brought into his house the *black* stone enveloped in a *white* silk and locked it in a chest, a symbolic gesture describing him as the "keeper" of the tradition (see Gaudefroy-Demombynes 27-29); Zubayr's new Kaaba, like Herod's new Temple, was a way to impose his supreme authority. Yet, as Ibn 'Arabî stated, Ibn Zubayr was wrong in doing so, since it is impossible to return to an exhausted and past cycle. In conformity with the cyclic moment, the Islamic Kaaba was veiled and cubic, the *hijr* being left outside (the Nordic side), while the Abrahamic Kaaba had a semicircular shape. The Abrahamic Kaaba appears oriented along the symmetry axis of the *hijr* (North – North-West), direction that represents Abraham's journey to Mecca, and, curiously, this direction is also that of Jerusalem (the change of *qiblah* from Jerusalem to the House of Allâh indicates a readjustment of the tradition).

[2] Gilis, *Pèlerinage*, p. 109.

[3] Michel Vâlsan saw this "unveiling" of the Kaaba in Dante's sayings: "Whilst eagerly I fix on him my gaze,/ He eyed me, with his hands laid his breast bare,/ And cried, Now mark how I do rip me: lo!/ How is Mohammed mangled" (*Inferno* XXVIII).

Earth is composed vertically by a cubic base and a semispherical dome, but also horizontally by a rectangular nave and a semicircular apse (oriented towards the East in the case of the Christian church); this is also the complete form of a Masonic temple, where to the "long square" (the *Hekal*, the Holy) is added the *Debir* (the Holy of Holies) in a semicircular shape; in a mosque, the *mihrâb* corresponds to the apse of a church.[1] Even though the communication of the secondary centers with the Center is completely invisible, in the last stages of the *Kali yuga* the "orientation" is more visible, at least in its symbolic illustration, like the "orientation" of the church towards the East and the "orientation" of all mosques towards Mecca, which is a reflection of the unique Center.[2]

No doubt, for Islam, Mecca is the unique center of the world and the Kaaba is the center of Mecca and a substitute for the "True House" (station Arafa)[3]: "Lo! the first Sanctuary appointed for mankind was that at Becca, a blessed place, a guidance to the peoples."[4] Mecca is the "mother of all cities" (*Umm al-Qurâ*), the "navel of the earth" (*surratu-l-ard*),[5] the first

---

[1] *Symboles fondamentaux*, p. 264, Burckhardt, *Principes*, p. 164.
[2] See Guénon, *Aperçus sur l'ésotérisme islamique*, p. 32.
[3] Arafa is the center as temple, while the Kaaba is the temple as center (the absolute center is *Lâ Maqâma*, the Non-station, which Gilis related to "the visit to Medina"; see *Pèlerinage*, pp. 311-320). "The pilgrimage is Arafa" (*al-Hajj 'Arafa*) says a *hadîth* mentioned by Michel Vâlsan, who added that the essence of the Islamic pilgrimage represents this phase (Études Traditionnelles, no. 404, 1967, p. 250). Gilis, after Vâlsan, stressed the importance of Arafa as center. Arafa is the apex of the pilgrimage, followed by the descent from Arafa to Mecca (*ifâda*), which Gilis compares with a "descendant realization" (and we know how important this phase is for Ibn 'Arabî). Arafa is the representation of what Guénon calls "the center of the integral being," whereas the Kaaba designates the center of the human state (Gilis, *Pèlerinage*, pp. 16-17, 24, 39).
[4] *Qur'an* 3:96. Some considered that Becca instead of Mecca was an orthographic error; in fact, there is no error, since the two letters B and M play an important symbolic role, related to manifestation and non-manifestation.
[5] Gaudefroy-Demombynes 30.

affirmed point (as Michel Vâlsan specified),[1] the point beneath the letter *bâ*, called *bakkah*,[2] and the heart of the believer.[3]

For the Jews, the Temple of Jerusalem was also the unique center, since God was one, and God ordered "all men to rise up, even from the furthest boundaries of the earth, and to come to this temple." However, as Guénon said, Islam, as the most recent traditional form, had to express the affirmation of Unity in the strongest way possible and with the highest persistence. This essential reality of Uniqueness and Unity became central in the Prophet Muhammad's teachings, and it is easy to discern how the new tradition, as the last revealed one, strove to avoid the errors of the previous religions and fight against any type of idolatry,[4] including the spiritual concept of the "son of God."[5] The moment of the cycle imposed such an explicit affirmation of Unity, but also imposed a "violent" manifestation, since the Prophet, learning from Jesus' terrestrial life, incessantly battled against those who considered him an impostor and the *Qur'an* a

---

[1] Gilis, *Pèlerinage*, p. 60.

[2] *Bakkah* is the birth into the sphere and *makkah* is the death into the cube. See also Gilis, *Pèlerinage*, p. 61.

[3] For Ibn 'Arabî, the Kaaba is only the exterior center, while the heart is the invisible, inner and more real center, the generous house (*bayt karîm*) and the immense sacred territory (*haram 'azîm*). On the other hand, the visible Kaaba is just one of the Houses, since each heaven has its own center or house, similar to Kaaba and to the Visited House (*al Bayt al-Ma'mûr*). See also *The Meccan Revelations*, I, p. 189.

[4] Having in view the Arabs' strong attachment to idols, the Prophet Muhammad could not accept anything but Unity and Unicity, rejecting therefore the Christian Trinity ("They surely are infidels who say: Lo! Allâh is the third of three; when there is no God save the One God," *Qur'an* 5:73) and the Israelites' errors. "Whoso ascribeth partners to Allâh, he hath indeed invented a tremendous sin" (*Qur'an* 4:48); "Hast thou not seen those unto whom a portion of the Scripture hath been given, how they believe in idols and false deities" (4:51); "Verily, God forgives not associating aught with Him, but He pardons anything short of that, to whomsoever He will" (4:116).

[5] "And the Jews say: Ezra is the son of Allâh, and the Christians say: The Messiah is the son of Allâh" (*Qur'an* 9:30). See also Titus Burckhardt's commentaries in his article *De la Thora, de l'Evangile et du Qorân* (Études Traditionnelles, no. 224-225, 1938).

forgery[1] or just fiction[2]; "For Jesus himself testified, that a prophet hath no honour in his own country,"[3] yet Prophet Muhammad's mission was not to be "defeated" as Jesus was by his countrymen,[4] but to fight, and he revealed himself as the true Messenger.[5]

---

[1] "O ye who believe! Believe in Allâh and His messenger and the Scripture which He hath revealed unto His messenger, and the Scripture which He revealed aforetime. Who so disbelieveth in Allâh and His angels and His scriptures and His messengers and the Last Day, he verily hath wandered far astray. Lo! those who believe, then disbelieve and then (again) believe, then disbelieve, and then increase in disbelief, Allâh will never pardon them, nor will He guide them unto a way [the Jews who believed in Moses and then in the golden calf, and rejected Jesus]" (*Qur'an* 4:136-137). "The people of Noah said the God's messengers were liars" (26:105). "O ye unto whom the Scripture hath been given! Believe in what We have revealed" (4:47). The Prophet, fighting to demonstrate his genuine mission, even insisted that the Jews and the Christians have hidden the prophecy that should be in their Scriptures regarding the coming of Muhammad. "Do ye not see those who have been given a portion of the Book? they buy error, and they wish that ye may err from the way!" (*Qur'an* 4:44); "Some of those who are Jews change words from their context and say: We hear and disobey" (4:46). See also Titus Burckhardt's commentaries in the mentioned article.

[2] "Nay, say they, (the Qur'an is but) muddled dreams; nay, he hath but invented it; nay, he is but a poet" (*Qur'an* 21:5); "This is naught but a lie that he hath invented, and other folk have helped him with it, so that they have produced a slander and a lie" (25:4). The reference to the people who helped could be about the Nestorian Sergius and the Jew Ibn Salâm. "Or say they (again): He hath invented it? Say: If I have invented it, upon me be my crimes, but I am innocent of (all) that ye commit" (*Qur'an* 11:35).

[3] *John* 4:44.

[4] Or other Jewish prophets: "And never came there unto them a messenger but they did mock him" (*Qur'an* 15:11).

[5] In the Christian tradition, Jesus had to be crucified, since this was, in essence, the profound reason why He came to earth. The Prophet Muhammad's perspective was in accord with his mission, different from Christ's paradigmatic earthly existence. We cannot and must not compare the various traditions and religions. "Verily, whether it be of those who believe, or those who are Jews or Christians or Sabæans, whosoever believe in God and the last day and act aright, they have their reward at their Lord's hand, and there is no fear for them, nor shall they grieve" (*Qur'an* 2:62). See also Nicolas de Cues, *La paix de la foi*, where Cusanus explained how God sent to different nations different prophets and teachers, and how serious conflicts developed when each community opposed its religion to another (p. 32); for Cusanus,

However, after Prophet Muhammad's death, this unity was soon spoiled by the appearance of so many sects and groups, which were the product of the human factor, that the acknowledgment of the Unicity became all the more urgent and essential. Even Mecca, which generally remained the unique center, was not spared, and in the year 929 the Karmatians pillaged the Kaaba, thirty thousand pilgrims, it is said, were slained, and the black stone was abducted.[1] The Karmatians hoped, with good reason, that the seizure of the black stone, that is, of the Tradition, would help their own center to prevail,[2] but in 952 they returned it of their own accord.[3] This end of the first Christian millennium was turbulent in more than one way, and we witness the rise of the Omayyad caliphate in Spain (from the year 929), and of the Fatimid caliphate (from the year 909), challenging the Abassids, who have moved Islam's center of power even farther from Mecca, to Baghdad. For their part, the Karmatians, initially connected to the Fatimids, attempted to establish a paradisiacal society[4] replacing the Islamic one (which lost the black stone), yet again, a society based upon acute violence.[5]

---

there is only one religion in a diversity of rites (p. 33) (Cusanus' expression "unity of religion," p. 39, was used later by Schuon).

[1] Similarly, the Philistines abducted the Ark of the Covenant from Shiloh.

[2] A tradition says that "Abu Tahir suspended the black stone to the seventh pillar of Kûfah's mosque, hoping that the *hajj* will be transferred here from Mecca; later, in 929, the Karmatians moved the stone to Hajar" (Gaudefroy-Demombynes 49).

[3] The abduction of the black stone, we may note, was a sign predicting a change of cycles and, in fact, the Karmatians themselves considered the imminent commencement of a new cycle. The Karmatian evilness was present even when the stone was returned: it is said that the messenger of the Karmatians tried to plant a terrible doubt in the minds of the Meccans, telling them that maybe this is not the genuine stone; yet, since the black stone (like the pillars of Enoch, see Josephus) was immune to fire and water, the Meccans could prove that it was the real thing (Gaudefroy-Demombynes 50).

[4] The Karmatians possessed an "initiation" in seven steps.

[5] We may assume that God punished Islam through the Karmatians, since "Unto Allâh belongeth whatsoever is in the heavens and whatsoever is in the earth. He forgiveth whom He will, and punisheth whom He will. Allah is

These tumultuous times for Mecca made Jerusalem flourish again as a center; nonetheless, the signs of the Millennium's end were also present there, preparing for the Christian Crusades.[1] The law of the cosmic cycles made the Abbasids in Baghdad decline and even though the Abbasid Caliph kept the title and the apparent power, the actual rulers were, first, the Bûyid dynasty, and then the Seljuk Turks, who captured Baghdad in 1055 and Jerusalem in 1076 (from the Fatimids).[2] The Fatimids recaptured Jerusalem in 1098, and we could say that it was God's Will to use the Seljuk Turks and the Fatimids as catalysers for this unprecedented Christian pilgrimage – the Crusades, which were, in their profound significance, nothing less but the "quest of the Center."[3] In 1099, the Crusaders conquered Jerusalem; the cross was installed crowning the Dome of the Rock, which became, for 80 years, *Templum Domini*. The Dome of the Rock was given to the Augustinians, while the *Al-Aqsa* Mosque was turned into a royal palace by Baldwin I in 1104. Beyond the human factor and the historical circumstances, we should not forget the essence of these events, which is the perpetuation of the Center, since with the

---

Forgiving, Merciful" (*Qur'an* 3:129); "Allâh doeth what He will" (*Qur'an* 2:253).

[1] At the end of the millennium, the Christian churches in Jerusalem were ruined and despoiled, the patriarch burnt alive.

[2] We see, in different garments, the old rivalry between Egypt and Assyria, with Jerusalem in the middle.

[3] In Masonry, the same "quest of the center" is illustrated by the return of the Jews and the building of the second temple. Regarding God's Will, we have described already how God's Mercy and God's punishment worked cyclically in the case of the Jews, which the Judaic tradition acknowledges. Consequently, we could apply the same scenario for Jerusalem during the Muslim conquest: in the 8th Century, a severe earthquake destroyed part of *Al-Haram El-Sharif*; in the 9th Century, the plague of locusts and another significant earthquake occurred; in 1016, a new earthquake caused the great cupola of the Dome of the Rock to fall down, all these being God's signs, reminding people about the Uniqueness and fidelity; in 1060, marking the Millennium change and being viewed as a terrible sign, the great candelabra suspended from the dome fell upon the Rock. "We sent against them [Pharaoh] the flood and the locusts and the vermin and the frogs and the blood – a succession of clear signs" (*Qur'an* 7:133).

Crusades, Jerusalem and the Temple Mount were once again recognized as center, a center that did not depend on which of the three Abrahamic religions was governing the holy place. In 1119, the Order of the Temple was founded as a living proof that the esoteric meaning of Solomon's Temple was not at all something obsolete and forgotten.[1] The Knights Templar, who considered the Dome of the Rock to be near the ruins of the Temple of Solomon, made their headquarters in the Temple Mount (in the area of the *Al-Aqsa* Mosque), adjacent to the Dome, which became the church of the Order of the Temple and it was featured on the official seals of the Order's Grand Masters[2]; the Dome of the Rock also became the architectural model for Templar churches across Europe.[3]

The attraction to the center was so powerful and indomitable (almost chaotically) that in the 9th Century the Church had to introduce a regulation through which it granted or withheld the privilege of pilgrimage; and later, in the time of the Crusades: "And when they heard the name of Jerusalem, the Christians could not prevent themselves, in the fervour of their devotion, from shedding tears; they fell on their faces to the ground, glorifying and adoring God, who, in His goodness, had heard the prayers of His people, and had granted them, according to their desires, to arrive at this most sacred place, the object of all their hopes." The attraction of the center made them come and die in hundreds of thousand; it was "a motive power strong enough to enable them to endure hardships and privations almost incredible, and to combat with forces

---

[1] We cannot emphasize enough that the Knights Templar's "interest" in the Temple of Solomon had nothing to do with some "moralizing opportunities," or with some secret documents or treasure buried there, or with some allegories regarding the building of the Temple; their "interest" was purely spiritual and it was related to the symbolism of the center and to the very "technical" and "positive" rules observed by the spiritual influences.

[2] Corbin 371.

[3] See the octagonal church in Tomar, Portugal, the octagonal church of the True Cross in Segovia, Spain, and the Church of Saint Mary of Eunate (located near Muruzábal, Navarre, Spain). The last one is on the famous Way of St. James of Compostela and has an octagonal plan as well.

numerically, at least, ten times their superior." And even though the human factor produced a series of disharmonies (quarrels, arrogance, looting, lust, greed, sinful priests),[1] there always was a right intention and a sacred orientation aiming at Christ and the Center, and there always were among the multitude some "righteous men." With the conquest of Jerusalem, the Crusaders became natives of the center; Foulcher of Chartres narrated: "Consider how the West has been turned into the East; how he who was Roman or Frank has become here a Galilean or an inhabitant of Palestine; he who was a citizen of Rheims or of Chartres has become a citizen of Tyre or of Antioch ... one has married a woman who is not of his own country – a Syrian, an Armenian, or even a Saracen who has abjured her faith ... they all talk different languages, and yet succeed in understanding each other ... The stranger has become the native, the pilgrim the resident."

---

[1] Robert of Normandy lamented: "Miserable men that we are! God judges us unworthy to enter into the Holy City, and worship at the tomb of His Son."

www.ingramcontent.com/pod-product-compliance
Lightning Source LLC
Chambersburg PA
CBHW070309240426
43663CB00039BA/2511